A HISTORY OF
GEORGIA AGRICULTURE
1732-1860

JAMES C. BONNER

A History of Georgia Agriculture 1732-1860

UNIVERSITY OF GEORGIA PRESS   Athens

To
WARREN, PAGE, JIM, AND ALLEN

Paperback edition, 2009
© 1964 by the University of Georgia Press
Athens, Georgia 30602
www.ugapress.org
All rights reserved
Printed digitally in the United States of America

The Library of Congress has cataloged the hardcover edition of this book as follows:
Library of Congress Cataloging-in-Publication Data

Bonner, James Calvin.
A history of Georgia agriculture, 1732–1860 [by] James C. Bonner.
viii, 242 p. 25 cm.
Bibliography: p. 233–236.
1. Agriculture—Georgia—History. I. Georgia agriculture, 1732–1860. I. Title.
S451.G3 B6
630.9758   64-22780

Paperback ISBN-13: 978-0-8203-3500-1
ISBN-10: 0-8203-3500-2

Contents

	Preface	vii
I	TIDEWATER LAND AND COLONIAL LABOR	1
II	COLONIAL AGRICULTURE	13
III	THE PASSING OF THE FRONTIER	32
IV	THE RISE OF UPLAND COTTON	47
V	SOIL EXHAUSTION AND EMIGRATION	61
VI	DIVERSIFICATION AND ECONOMIC INDEPENDENCE	73
VII	SOIL CONSERVATION AND SOUTHERN NATIONALISM	93
VIII	AGENCIES FOR PROMOTING AGRICULTURE	110
IX	THE QUEST FOR GRASSES AND IMPROVED LIVESTOCK	127
X	ORCHARDS AND VINEYARDS	149
XI	GARDENS AND BUILDINGS	168
XII	COTTON, CORN, AND SLAVERY	186
	Notes	204
	Selected Bibliography	233
	Index	237

Preface

This study was first suggested to me by Professor Fletcher M. Green of the University of North Carolina, under whose direction I completed in 1943 a Ph.D. dissertation on the agricultural reform movement in Georgia. That phase of the present work was made possible by a grant in 1941 by the Institute for Research in the Social Sciences of the University of North Carolina. Special acknowledgment must be given to Professor Green for his encouragement and skillful assistance in the earliest phases of this task.

The completion of the full story of Georgia's ante-bellum agricultural history was aided by Duke University which extended to me as a visiting scholar the hospitality of that campus in the summer of 1953 and by a grant from the Southern Fellowships Fund for the summer of 1958. The latter was used principally to study in detail the early agricultural inventories in selected Georgia counties. Portions of two articles, one of which I published in *Agricultural History* in January 1945, and the other in *The Georgia Review* in 1963, have been reproduced here by permission of the editors of these journals.

Many individuals have aided the undertaking, each in a small but highly significant way. Foremost among these are Dean Donald H. MacMahon of the Woman's College of Georgia, former President Henry King Stanford, and his successor, Robert E. Lee, of the same institution, all of whom from time to time have afforded me some relief from departmental duties without which the completion of the study would have been impossible. The Committee on Faculty Research at this college has provided funds for the clerical and stenographic aid of several of my students, among whom are Miss Carol Walton, Miss Sybil Kelly, Miss Frankie

Shirey, and Miss Irene Porter. The librarian of the Woman's College of Georgia, Miss Virginia Satterfield, has served me with efficiency and courtesy, as have other librarians and archivists throughout the South. I owe a special debt of gratitude to Professor Spencer Bidwell King of Mercer University and to Professor Kenneth Coleman of the University of Georgia, both of whom read the entire manuscript. They offered pertinent suggestions based upon their scholarly knowledge of Georgia's past.

JAMES C. BONNER
Woman's College of Georgia

I

Tidewater Land and Colonial Labor

THE OLDEST agricultural region of Georgia was defined by the original Indian cession made in 1733. This grant by the Yamacraws was a narrow belt along the coast, less than thirty miles wide, from Savannah to the Altamaha River, including the chain of sea islands. After 1763 this area was extended southward to the St. Marys River, making a total length of 126 miles. The agricultural development of the tidewater lands and sea islands along the coast cannot easily be articulated with that of the upcountry settlements reaching from Ebenezer northward to the foothills above Augusta. With its population concentrated in such towns and villages as Savannah, Ebenezer, Midway, and Frederica, the lower tidewater region was the first to develop. Its early history is characterized by considerable experimentation in the growing of semi-tropical plants which often resulted in failure, in a method of land tenure ill calculated to succeed in a frontier environment, and in a system of labor which proved unsatisfactory for colonial enterprise.

The most productive land was that found on the sea islands, whose soils, a mixture of sand and black mould, yielded good crops of corn, indigo, potatoes, and vegetables. By the end of the colonial period they had begun to produce the most valuable cotton known to the markets of the world, and the land commanded high prices. These islands in their natural state were covered with pine, liveoak, hickory, and red cedar, and they were separated from the mainland by navigable creeks which provided an efficient inland navigation. Between the islands and the mainland was an expanse of salt marsh about five miles in breadth. This area was covered with a coarse grass which cattle did not relish but which offered a refuge for an endless variety of waterfowl.[1]

Back of these brackish marshes was the mainland where the soils were of uneven quality. Here were the swamp lands on which pines, oaks, and magnolias grew, and they were much admired by European settlers. The cypress trees which abounded in the undrained swamps provided a valuable source of shingles and barrel staves, and other types of trees were useful in the processing of lumber. Some twenty miles back of the salt marshes of the rivers great quantities of rice were produced on lands considered the most valuable in the province. The rice plantation came to be the most picturesque institution of the colonial period.[2]

Inferior in quality to the rice lands for agricultural use were those lands nearer the coast, situated between the tidewater streams. Originally covered largely with pine, wild grass, and small reeds, they afforded good range for livestock. Interspersed here and there were low ridges or "hammocks," on which grew oak, hickory, gum, walnut, and other hardwoods. After being cleared with simple tools they produced good crops of corn, potatoes, and indigo.[3] Georgians early discovered the inferior quality of the sandy soils on which they found only pine trees growing, with perhaps a sprinkling of chinquapin and hurtleberry shrubs. Known as "pine barrens," these extensive lands came to be avoided for general cultivation, although they would produce peach and mulberry trees, and the pines were valuable in the forest products industries. Much of the early discontent of the Georgia colonists stemmed from the fact that these barren soils often made up the major portion of their holdings.[4]

The arduous task of bringing into cultivation such a frontier as the Georgia coast presented in the first half of the eighteenth century required an adequate supply of labor. Because slavery was in conflict with the type of colony which the Trustees had in mind for Georgia the use of Negroes was prohibited. The Trustees envisioned Georgia as a community of small, self-sufficient landholders. Indentured white servants were to compose the principal labor force, but these in time would become freeholders and masters of their own farms. Therefore, the problem of labor supply remained persistent throughout the Colonial era.

The early English settlers of Georgia were unacquainted with the arts of agriculture. They represented a wide variety of urban trades and occupations, including writers, potash makers, wigmakers, mercers, and upholsterers, with only a sprinkling of skills related to the cultivation of the soil. The German Salzburgers, who proved to be the most successful husbandmen of the Colonial period, came largely from an urban mining community where they had been engaged in such occupations as carpentry, brazing, and toy-making.[5] The Portuguese Jews, of whom there were a relatively large number, tended to live in the towns and to avoid the exertions of agricultural life. The Scotch, whose immigration began in 1734, were highly regarded by General Oglethorpe for their fighting qualities and their sturdy habits, rather than for their agrarian skills. As a result they were settled on the southern frontier where their main function was to do garrison duty in the colony's defenses against the Spaniards.[6]

The Georgia Trustees brought over indentured servants not only for the use of independent colonists but for labor in public works under the supervision of Trust agents. In general the independent colonist in Georgia could not afford to pay the passage money in exchange for a servant's indenture, and when he did so he was likely to be left without means to provide himself with adequate clothing and subsistence.[7] This situation, together with the debilitating climate, decreased the effectiveness of the plan to use indentured servants as laborers.

The Trust servants, who were employed in various enterprises, not all of which were agricultural, generally received better treatment than those who worked for the planters, largely because the former were less efficiently supervised. Many were engaged in constructing roads, forts, and public buildings, sawing timber, and operating public grist mills. The largest number was used in raising food on the public domain for their own subsistence and for the subsistence of charity colonists. The largest unit of these was Bouverie Farm at Savannah, where, in 1739, the Trust maintained sixty servants at a total annual cost of £730.[8] Employment of Trust servants in public farm operations generally was disappointing in its results, for often they failed to produce enough for even their own subsistence. Finally, in 1743, the Trustees discontinued the employment of servants in all public works, but they continued to finance their transportation to the colony, where their indentures sold at a reasonable price to individual planters.[9] The growing liberality of the indentures is illustrated by the action of the

Common Council of the Trustees in 1749, in providing for the passage of over seventy German Protestants. Their apprenticeship was limited to four years under certain specified masters, and they were allowed to work for themselves one day each week or else to have 205 days deducted from their total term of service. Upon release they were to receive twenty acres of land, the same quantity as stipulated for other servants.[10] By this time the Trustees had abandoned the earlier practice of recruiting servants from the streets of London, and were turning to rural areas for a supply of youths inured to hard labor. Many were recruited from the highlands of Scotland and Switzerland, and from Germany. The German servants came to be held in high esteem, despite the language barrier.[11]

The growing efficiency of white servants, however, did not end the complaint of many colonists against the prohibition on Negroes. Criticism was greatest among the English settlers in the Savannah area where it was claimed that the climate was wholly uncongenial to white laborers in the hot and humid swamplands. The introduction of Negroes even as common hired servants was forbidden in 1741 when it was ruled that free blacks could not settle in the colony.[12] During the following year Thomas Stephens went to England where he presented the petition of the disgruntled planters seeking relief from the prohibition against slavery. "The poor people of Georgia may as well think of becoming Negroes themselves (from whose Condition at present they seem not to be far removed)," he wrote, "as of hoping to be ever able to live without them. . . ."[13]

The colonists were by no means unanimous in their opposition to this policy of the Trustees. The Salzburgers living at Ebenezer were opposed to the introduction of slaves, largely on social and religious grounds. Under the leadership of their minister, John Martin Bolzius, they argued that the presence of Negroes would lower wages and living standards of white servants, who would be forced to take the more remote and undesirable lands. Therefore the poorer classes would be discouraged and obliged "to seek their Livelihood in the Garrisons, Forts, Scout Boats, Trading Boats," wrote Bolzius, "or to be imployed amongst the Negroes upon a Gentleman's Plantation, or they are forced to take Negroes upon Credit of which they will find . . . Sad Consequence on account of their debts."[14] The Highlanders at Darien expressed opposition for military reasons.[15] They stated that their nearness to the Spaniards, who had proclaimed freedom to all runaway slaves, made it impracticable to keep them, and they also noted that slavery was "shocking to human Nature."[16]

Despite the Trustees' prohibition, a few slaves from South Carolina were used clandestinely on remote back-country plantations, and when apprehended they were seized by the magistrates and sold. By 1748 it was reported that the law had become so generally disregarded that slavery was fairly general. Finally, on August 8, 1750, the law was repealed.[17]

The great majority of Colonial Georgia's slaves came from South Carolina and the older colonies, whence emigrating planters brought them after 1750. A few were brought from the West Indies by merchant traders who had entered the Georgia slave market. The great bulk of these slaves were used on the rice and indigo plantations in the coastal parishes. Crops grown in the region around Augusta were ill suited to slave labor. Emigrants to that region before the end of the Revolution came largely from the Carolina upcountry, and they brought with them a tradition of subsistence economy. They employed relatively few slaves until the beginning of the nineteenth century when upland cotton came into prominence.[18]

The land system decreed by the Georgia Trustees was as great a source of discontent as its labor system. Each was inspired as much by military considerations as by either philanthrophic or economic designs. While the charter allowed grants of 500 acres, the Trustees reduced this figure to fifty acres each for settlers coming over at the expense of the Trust. These together with the indentured servants formed a large majority of the settlers arriving before 1750. The full limit of 500 acres was granted only to those settlers who came at their own expense and who brought with them six servants, a figure later increased to ten.[19]

Not only were the land grants relatively small in size, they were also restricted as to use, and the title itself was in *tale male*. Land could not be sold or mortgaged by the grantee. Since women were not expected to serve in the militia, they could not acquire title to land in any manner, except by special permission of the Trustees. Among the nine conditions under which the holder might have to forfeit his grant was the failure to grow mulberry trees, essential to the production of silk, which the Trustees were determined to promote. Somewhat like the tax on realty was a quit rent of four shillings per hundred acres, to be collected by the Crown after a ten-year occupancy. Later the rate was increased to twenty shillings on the claim that additional income was needed to finance local administration. However, because of the precarious economic condition of the colonists and their opposition to quit rents, none was collected before 1750.[20]

The Georgia holding was in fact a military fief and each male inhabitant was to be a soldier as well as a planter. This is illustrated by numerous regulations. The equipment issued to each charity colonist after 1735, for example, consisted of a "watch coat," musket, and bayonet, along with a spade and two hoes, a hatchet, hammer, handsaw, gimlet, drawing knife, a pair of hooks, and a frying pan. For a year's maintenance he received a soldier's subsistence ration consisting of 300 pounds of beef or pork, 114 pounds each of rice, flour, and peas, and also portions of salt, sugar, butter, vinegar, lamp oil, spun cotton, and soap.[21] Since most of these early Georgians were products of urban communities, they required training both in the art of cultivating the soil and in the use of their military accoutrements.[22]

Not only was military service a requirement of tenure, but the land holdings were located in such a manner as to promote compact settlements, giving each town or village the character of a military garrison. The town lots were laid off in checkerboard fashion with each holder assigned an eighth of an acre for his dwelling. On the immediate periphery of the town was a garden plot of four and seven-eighths acres for each householder. Some distance farther lay the remainder of his holding, consisting of forty-five acres. The last was often in the shape of a long triangle with its apex nearest the town boundary, thus increasing the total length of the enclosure to the inconvenience of the holder.[23] The towns were further divided into tythings, and the old Saxon custom of appointing tythingmen for each 500 acres was followed. Ten freeholders whose houses were contiguous formed a tything, a term applied both to the unit of town lots and to the larger agricultural division. The full size of a complete tything in Georgia was 640 acres, comprising twelve lots, two of which were reserved by the Trust for public uses, and an additional forty acres for roads.[24]

All of these land units were surveyed prior to settlement and assigned a number. Only the non-charity colonist, entitled to the maximum of 500 acres, often obtained a plat which was surveyed indiscriminately so that its boundaries were constructed in an irregular manner. Hence his enclosure was more likely to contain only the choicest land. On the other hand, the poor and uneven quality of the smaller plats was a constant source of discontent among other settlers.[25] The servants upon release from their indentures frequently complained that the better lands were engrossed by those who received the first assignments. It seems probable that the servant released from indenture, who drew only twenty acres, obtained the poorest land.[26] This situation was remedied somewhat in 1741 when the Common Council authorized additional acreages for the smaller freeholders having only pine barren land.[27] To further remove complaints against discriminatory treatment in the assignment of land, a lottery system was used at Vernonburg during the following year for the distribution of holdings to twenty-six German servants whose indentures had expired. A map of the farm plats was made and numbers assigned to them and to each corresponding town lot. After the numbers were drawn the titles were recorded, and each new freeholder was paid twenty shillings for the purchase of tools. The drawing of the numbers was celebrated by the drinking of wine and rum.[28] This was perhaps the first use in Georgia of the lottery system which came to characterize the state's policy of land distributions at the beginning of the eighteenth century.

After 1752 the plats granted to all settlers began to shift toward the irregular pattern, indicating a more serious official regard for the quality of land being granted. For example, a 400-acre plat surveyed for Benjamin Bakers in 1761 contained a total of eleven sides, only two of which were parallel.[29] Even more significant was the change in the nature of ownership. Possibly no feature of Trustee policy was more highly criticized than that of granting only *tale male* titles. In December 1738, the Trustees were petitioned by 121 people at Savannah to give a fee simple title to their lands, which, if granted, they said, "would both occasion great Numbers of new Settlers to come amongst us, and likewise encourage those who remain here, cheerfully to proceed in making further improvements as well as to retrieve their sunk Fortunes, as to make provision for their Posterity."[30] Thomas Coram, a member of the Trust, agreed that *tale male* tenure deprived "the Females of their just Right given them by God and Nature." More significant, however, was his liberal concept of the importance of giving settlers full ownership of their property, thus to enhance land values in a frontier society. "Land is of no value now, nor ever can be," he said, "until cultivated and improved by their great Labor and Expense." He believed that a greater profit incentive would encourage the colony's growth.[31]

Subsequently the Trustees altered their policy by making any descendant of a grantee, including women, or any other person, capable of "enjoying by Devise of Inheritance" any quantity of land up to 2,000 acres.[32] Still cognizant of the military nature of the colony, however, the Trustees continued to discourage large holdings and isolated settlements. It was not until 1750 that the title became completely a fee simple one. In that year all grants were extended to "an absolute Inheritance, the Grantees, their Heirs and Assigns [forever]."[33] After the passage of Florida from Spanish to British hands in 1763, Georgia rapidly lost its last vestiges of a military colony.

The change to a fee simple title and the removal of the prohibition against the use of Negro slaves both came in 1750, and were followed in 1752 by the end of the trusteeship and the inauguration of the Royal Province of Georgia. These changes encouraged the rapid rise of large plantations devoted to rice and indigo. The plantations were confined largely to the sea islands and to the Savannah and Altamaha regions, and represented an extension of the system already in use in South Carolina. The expansion of this system continued rapidly throughout the Colonial period. By 1773 sixty men each owned 2,500 or more acres. Of these twenty owned more than 5,000 acres and possibly three held in excess of 20,000. Under the land laws prevailing at that time each grantee received above a family right an additional fifty acres for each slave. The total number of slaves in the hands of the sixty largest planters probably approached 7,000, or more than half the total of all slaves in the province.[34]

Under Georgia's royal governors, who now had control of land grants, the province experienced a throbbing economic growth and an economic prosperity undreamed of in the days of the Trust. In 1763, and again ten years later, Indian land cessions extended the original boundary line approximately five miles westward, and in addition embraced the corridor between the Ogeechee and the Savannah rivers northward to the Little River above Augusta. Another tract was ceded along the Atlantic coast from the Altamaha south to the St. Marys. This cession, known as the trans-Altamaha region, was confirmed in 1765. In the treaty of 1773, the Creeks and Cherokees together ceded a large area above Augusta north of the Broad River in the settlement of some

£40,000 in debts owed by the Indians to British traders. Originally known as the "New Purchase" and containing some 1,500,000 acres, it was organized in 1777 as Wilkes County.[35] In addition to the "New Purchase," the Creeks alone ceded in 1773 another wide area in the South between the Savannah and the Altamaha rivers, to become the last treaty of cession made under British auspices. All of these treaties were negotiated under the royal administration of Governor James Wright, and they increased by more than fivefold the land area which had been subject to Trust administration.[36]

Thus the map of Colonial Georgia had come to resemble a thick L-shaped figure with Savannah at the apex, and prongs extending southward to the St. Marys and northward beyond Augusta. Each prong of this figure came to represent a separate agricultural region. In 1764, soon after the St. Marys became the southern boundary of Georgia, the King extended the western limits of the province to the Mississippi River. The southern portion of what later became Alabama and Mississippi, south of the thirty-first parallel, was not included in the limits of Georgia until after the Revolution.

The population of Georgia increased rapidly under the royal governors. From 1,700 whites and 420 Negroes in 1751, the number of inhabitants had increased to 33,000 people by 1776, with Negroes composing forty-five per cent of the total. Population continued to grow during the Revolution and by its end was estimated to have reached 50,000.[37] The great majority of those who came to Georgia after 1750 were immigrants from the older colonies, particularly Virginia and the Carolinas. Most of these entered the province by crossing the Savannah River in the vicinity of Augusta, and they settled in the New Purchase. The size of individual plantations began to grow perceptibly, although the amount of land under cultivation continued to be small. Theodore Keiffer, who began his career in Georgia as an indentured servant, in 1757 was the possessor of 620 acres and five slaves, and he was somewhat typical of the rising planter group.[38] Representing the modest planter was John Adam Treutlen, a Salzburger, who owned 1,300 acres and 23 slaves in 1769.[39] It is estimated that by 1773 five per cent of the landowners had received about twenty per cent of all lands granted by the royal governors and they also possessed over half of all Negro slaves in the province. In addition their holdings constituted the bulk of the better and more favorably situated lands. John Graham's holding of

26,000 acres was the largest in the colony, but those of Governor James Wright and James Habersham each exceeded 20,000 acres. The governor, whose property was soon to be confiscated by the Revolutionary government, probably was the wealthiest man in Georgia with 24,578 acres and 523 slaves.[40]

Since the royal governor, like the Trustees before him, had control of all land granting, there was always the possibility that he might abuse the power to enrich himself and to reward his friends. While there is no evidence that Wright was dishonest in the use of his prerogatives for personal gain, his policies tended to favor the interests of the wealthier planters and speculators over those of small farmers. It was largely in this manner that he was able to influence certain local political measures favorable to the Crown.

Beginning in 1760 the governor began granting 100 acres to every family head and an additional 50 acres for each family member, including slaves. Under this new "family rights" system only the better lands were surveyed and patented. In an attempt to acquire the best lands possible the grantee frequently had the plats surveyed in irregular patterns with wild lands of poorer quality interspersing the patented grants. On the unsurveyed lands squatters might settle unchallenged. Again in 1773, in order to encourage settlers from older colonies to move into the New Purchase, Governor Wright ordered the land surveyed into 100- and 1000-acre tracts and offered it to settlers on liberal terms, exempting the buyers from the payment of quit rents for a term of ten years. Settlers were also allowed to buy extra land in addition to what they had received under the headrights restriction. Wright also promised to build a fort in the territory and to garrison it against disturbances from "disorderly hunters, vagrants, and wanderers, and by straggling Indians."[41]

While the creation of a landed aristocracy was in keeping with British traditions, the favoring of conservative interests over those of small farmers appears to have originated somewhat as a local policy of the Crown's officials in the colony. Among the more affluent men in the community, these officials often acquired large tracts of land for speculative purposes. The Crown's view of the matter may be judged more accurately from its regulation of 1774 for disposing of ungranted lands in Georgia and in six other colonies. The regulation instituted greater royal control over land granting and ordered that lots be sold to the highest bidder rather than given away under the family rights system, a policy somewhat more favorable to non-resident speculators.[42] Opponents of this policy believed that it prevented the rapid settlement of the back country. This sentiment was reflected two years later in the Declaration of Independence by the statement that the King "has tried to prevent the population of these colonies."

With the coming of the Revolution in 1776 the control and distribution of public land came under the direction of the Whig faction, which group had been most at odds with British policies. Georgia's land situation was unique among the thirteen rebelling colonies, for it possessed at this time one of the larger public domains available for future settlement. In addition, Georgia's total area, including unceded Indian lands, consisted of a vast potential empire extending to the Mississippi River. Including its western lands, to which Indian claims had not yet been extinguished, over ninety-five per cent of the Georgia area was classified as public domain. Because of this fact land policy became the primary political concern of Georgians for a half century after the Revolution, helping to explain the highly transitory nature of its agricultural population throughout most of the nineteenth century.[43]

Of all the lands in Georgia secured by Indian treaties during the Colonial period, only about one-third had been patented at the beginning of the Revolution, leaving approximately five and a half million acres to which Indian claims had been extinguished.[44] With the exception of the New Purchase of Wilkes County, much of this land was in pine barrens and considered suitable only as grazing range. However, the Whig government in 1778 and again in 1782 passed acts confiscating property of Tories who were attainted for high treason, thus bringing into public possession approximately two hundred thousand acres of the best and most highly cultivated lands in the state, together with such slaves, livestock, and other chattels which had not been carried away.[45]

This large area in the public domain was Georgia's principal means of financing the Revolution. Land was the principal collateral against which the state issued paper currency, and offering soldier land bounties was the principal feature of war-time measures. Among the earliest of these was a proclamation by President Archibald Bulloch of the Provisional Assembly in July 1776, promising one hundred acres to all who would enter military service for three years. In recognition of the state's exposed Indian frontier, acts were passed in the next two successive years designed to entice emigrants from the older states to settle in the Wilkes

County area and to serve as minutemen.[46] A headright of 200 acres was offered with an additional fifty acres for each member of a family or for slaves up to ten in number. To encourage needed war-time industries a bonus of 100 acres was offered for each grist mill, 500 for each saw mill, and 2,000 acres for an iron furnace.[47] As the war progressed it became increasingly difficult to obtain enlistments. In 1781 tax exemption for ten years was offered in addition to a bounty of 250 acres. Finally, in 1782, an act directed the governor to issue land bounty certificates to anyone whose commanding officer certified as to any kind of military service in the war.[48]

Georgia's public domain after 1781 became the largest of any American state. In the hands of a weak and debt-ridden government, it became the target of great speculative interest after the Revolution and remained the central theme of Georgia history for the next fifty years.

II

Colonial Agriculture

THE CHARACTER of Georgia's early agriculture was a logical result of the British mercantile philosophy, wherein Colonial productions were directed toward the economic self-sufficiency of the whole empire. Crops such as rice and indigo were large-scale plantation crops requiring considerable capital and slave labor, and hence they were beyond the means of the earliest settlers. While the growing of tobacco was a family-sized enterprise, that commodity had become a drug on the market by the time of Oglethorpe's arrival.[1]

It was believed that the Georgia climate would produce exotic plants not grown elsewhere in the British possessions. To test this theory a "Trustee's Garden" was laid out at Savannah, comprising a ten-acre experimental plat having a wide variety of soils. In this garden were grown oranges, olives, apples, pears, figs, vines, pomegranates, cotton, coffee, tea, bamboo, and also palma christi and other medicinal plants.[2] After eight years of growth, the olive trees bore fruit which failed to mature, and cultivation of these trees was abandoned. Orange trees were more successful, and in 1770 more than 3,000 gallons of the juice alone were exported from the colony.[3] This fruit was grown on the Georgia coast for more than a century afterwards, although subject to the hazards of winter freezes. The crop finally succumbed to the low winter temperature.

Of all the experiments in exotic productions, growing silk appeared to offer the greatest promise. The leaves of mulberry trees were essential for feeding silkworms, but the plantings required relatively little cultivation after the first two years. Usually grown in cornfields at a distance of thirty feet, mulberry trees required little more than annual pruning. Planting of these trees being one of the initial conditions of land tenure, it also became in 1750 a requirement for holding the office of deputy in the Commons House of Assembly. Later, in order to so qualify, one had to show a production of fifteen pounds of silk annually for each fifty acres of land he possessed. He was required in addition to have at least one female member of his family instructed in the art of reeling silk.[4]

The requirements of silk production were so intricate as to lend to the industry the character of a manufacturing enterprise. The mulberry trees began to bud about the first of May at which time the silkworm eggs were placed in small boxes lined with paper, each containing approximately an ounce of eggs. When kept in a warm place, usually by the fireside, the eggs began to hatch within a few days; after hatching the worms were fed initially on lettuce leaves. Worms from an ounce of eggs might produce from five to ten pounds of silk, but they would require to be fed several hundred pounds of leaves to bring them to maturity.[5]

The leaves had to be gathered singly by careful hand-picking so as to avoid being bruised. Clean plucking of leaves from the branches would damage the tree and cause it to die.[6] Since one worm would consume its own weight in food each day, the gathering of leaves was a constant process during the feeding period. They could not be kept for more than forty-eight hours, and then only by being turned two or three times daily.

When the worms began to grow workers removed them from the boxes and placed them on racks or shelves, in a manner to avoid damaging them with the hands. During the stage of rack-feeding they had to be protected against rodents, cats, and birds, and against extreme temperatures.[7]

The total feeding time lasted about six weeks, or to the end of April, at which time the worms changed from a white to a transparent flesh color, signifying the advent of the spinning phase. At this stage clean dry branches of birch, vines, or briars, to which the worms could attach themselves, were placed against the racks. By the end of the fifth day, silk balls had been spun on these branches and had begun to harden. If disturbed during the spinning period the worms might stop short of a full spin, with a resulting loss of silk. Then the balls, or "cods" as they were now called, were removed, and the silk was ready for unwinding and reeling, a task which had to be accomplished within ten days to prevent the aurelia from eating through and destroying the fiber. However, this could be avoided by destroying the worms inside the cod with a moderate degree of oven heat.

In the unwinding of the silk the cods had to be placed in warm water to which progressively higher heat was applied until the silk threads became visible. Then the end of the thread was found by the use of good eyesight and a fine brush, whereupon the first few yards were discarded because of poor quality. The remaining thread was unwound and reeled by hand very much as one would re-wind a thread ball. Four threads were commonly unwound and reeled together, during which process knots and breaks were carefully avoided.[8]

The heavier cods were laid aside for seed. These were allowed to produce a new crop of worms which would begin to emerge in fifteen to twenty days and to transform themselves into moths. Prior to this time a selection was made of the male and female cods, the former being pointed at both ends and somewhat smaller in the center than the female. The male and female cods were then tied together in pairs with a needle and thread. When the moths emerged there followed a day of mating. Then the male was thrown away and the female allowed to produce her eggs, after which she died. The eggs were then wrapped in paper or linen and placed in an earthen pot where they remained throughout the summer. In the winter they were removed to a chest filled with woolen scraps and left until the spring hatching season.

The silkworm egg was about the size of a mustard seed and of a dark brown color. The worm or caterpillar was initially of a cream color, having six feet and ten holders. In this form it underwent four stages or "sicknesses," each lasting about three days, wherein it ceased all feeding, changed its skin, and at each stage became shorter and thicker of body. Its next change was from a caterpillar to a chrysalis, or aurelia, in the shape of a small plum, and it was during this stage that it did its spinning. The final stage, important only for breeding, was the moth.[9]

The production of silk was not a year-round family enterprise. For about two months it usually required the constant attendance of two persons, the hardest work coming about two weeks before the spinning season. By 1741 some growers were producing two crops of silk cods in one summer, thus extending the season's length to twenty-six weeks. The earlier crop weighed in at Savannah in May 1741 totaled 600 pounds, with each individual crop weighing from five to forty pounds. It required an average of 300 cods to produce a pound of raw silk.[10]

Winding off the silk and placing it on reels for shipment to market was a task which required highly trained and skillful specialists. As with most phases of the work, women were employed in this task. Using a simple machine, and with one child to assist her in directing the thread, a woman could hardly reel more than a pound of silk in a day.[11]

The silk industry was highly subsidized both by the British Parliament and by the local government of the colony. In 1747 the Common Council voted £200 to defray the cost of bringing to Georgia a model of a silk-winding machine used at Montalban in France.[12] Two years later the Trustees appropriated forty shillings sterling for each woman in the province who learned the art of silk-winding within one year, and the regulation continued through 1751. A bonus of five pounds was given to the first three Salzburg women who acquired the skill. Other grants were made to improve both the quality and quantity of cocoons. In 1751 £300 was appropriated for the erection of a new and more ample filature at Savannah and also one at Ebenezer.[13] The local silk bounty attracted many South Carolinians to the Georgia filatures where the growers were paid the Trustees' subsidy, causing the Trustees to issue an order that no silk grown in South Carolina could be reeled at any Georgia filature.[14]

The most successful silk growers in the colony were the Salzburgers living at Ebenezer. Their toy-making skill was applied to making reels, an art in which they became so skillful that one of their machines was sent to England to serve as a model. While most Georgians began to abandon the culture of silk after 1751, the Germans remained in the business almost twenty years longer. In 1766, a year in which they produced 20,000 pounds of cocoons, the bounty paid by the British government was reduced from three shillings per pound of cocoons to one-half that sum. Within the next three years the production of raw silk in Georgia fell to 290 pounds.[15] In 1769 Parliament placed a bounty of twenty-five per cent on raw silk imported from the colonies, and this stimulated production to an average of 450 pounds annually for the next few years. During the Revolution Ebenezer was devastated by British invaders; the silk subsidy disappeared with the state's independence, and the industry slowly died.[16] Some silk was produced in Georgia as late as 1790, and sporadic efforts were made to revive the industry throughout the nineteenth century, but they resulted in ultimate failure.

At an early date silk growing was supplanted in importance by the cultivation of rice, a crop grown in Georgia almost from the beginning of settlement, it being at one time used to provide winter feed for cattle.[17] The development of large plantations after 1750, following the inauguration of Negro slavery and new tenure laws, stimulated a rapid expansion of the crop. Rice had been grown since 1700, in South Carolina, whence Georgians derived their knowledge of the techniques of its culture. The market for rice was unsteady during the Colonial period as a result of trade restrictions growing out of European wars. South Carolinians raised 100,000 pounds in 1742 when the price dropped to only thirty shillings per hundredweight, thus causing a large part of the crop to remain unsold.[18] While it continued to be a drug on the market throughout the remainder of that decade, rice became the principal staple in the latter part of the Colonial period.

Rice production in Georgia was confined to the British planters on the coast where the ebb and flow of the ocean tides could be utilized in alternately flooding and draining the fields, using only salt-free river water in the process. The rice plantations, known as inland swamps, lay on the rivers and on the smaller streams flowing into them.[19] The business not only demanded tedious labor and skillful management, but also required a heavy outlay of capital. An estimated forty slaves and 200 acres of suitable swamp land, in addition to tools, equipment for cleaning and processing, and food for the upkeep of the workers for one year, were necessary to begin the enterprise.[20]

The planting of rice was begun around the latter part of March. It was planted with a hoe in shallow trenches eighteen inches apart. As the plants sprouted and began to grow, the earth had to be laid up to the roots of the plant and the entire field kept clear of weeds. The initial hoeing was begun about the first of May after which the field was cleared of trash and flooded to a depth of three or four inches. If slime or froth appeared, the water was run off on one ebb and taken in on the next flood. The water was kept on the field about seventeen days, or a little less if the sun and heat were intense. The water was then leaked off slowly and the field was allowed to dry for several days. At this point occurred the second hoeing, which consisted of loosening the soil and combing up the fallen plants with the fingers. After a few days, or around the first of July, came the third hoeing, during which the field again was cleaned of trash in preparation for the second flooding. Often the third hoeing and the second flooding were done concurrently. The fourth and final hoeing was done in the water of the second flooding, with the workers standing in mire almost knee-deep. At this stage the hoe was never actually used except on high ground. Grass and weeds which escaped previous hoeings had to be pulled out of the water by hand. If the plants became flaccid the depth of the flood was increased to provide them with support. The second flood was kept on the field until time to harvest.[21]

The reaping occurred at the end of August or in September, after which the heads were dried on the straw and stacked in the barnyard. Subsequent processing was tedious and painstaking. The rice was first threshed and then winnowed. It was next placed in mortars and beaten by hand with wooden pestles. Finally, to eliminate dust and broken grains, it was sieved through two successive screenings. It was then ready to be put into barrels and sent to market.

While the processing of rice was extremely laborious, it was less irksome than the field work required in growing it. The field workers, often stripped to the waist and always in bare feet, were subjected to malaria and snake bites. The scorching heat of the sun and the putrid effluvia from the stagnated water gave the work an unwholesome character.[22] It was such conditions as these which, in their petition against the slavery prohibition law, the colonists claimed that Englishmen could not endure.

Somewhat more attractive to the English settlers was the work required on indigo plantations. Having been introduced into South Carolina by Eliza Lucas, this crop was grown successfully there before 1700, when it began to be neglected in favor of rice. The abrupt decline in the price of rice in the 1740's resulted in a revival of indigo production. About this time Parliament placed a bounty on the product of sixpence per pound, which was reduced to fourpence in 1765 and remained at that figure until the bounty was withdrawn after the Revolution.[23]

The Georgia indigo industry dates from about 1740, when its possibilities were mentioned at a meeting of the Trustees. Robert Miller, a botanist, informed that body of a more favorable season in Georgia for this crop than prevailed in the more northern colonies.[24] Subsequent planting of indigo on the Georgia coast produced flourishing crops. By 1750 the crop was well established along the Ogeechee River and on the sea islands.[25] Unlike rice, it was a crop which lent itself to a diversified, family-scale enterprise. A visitor to Jekyl Island in August 1746 described the highly diversified farm of Major Horton, which was perhaps typical of the small plantations of the period. He observed, "a very Large Barnfull of Barley not inferior to ye Barley in England, about 20 Ton of Hay in one Stack, as spacious House & Fine Garden, a plow was going wth. Eight Horses, And above all I saw Eight Acres of Indigo of which he has made a good Quantity. . . ."[26]

The planting of indigo began about the end of March, when the seed (generally obtained from Guatemala) were placed in drills eighteen inches apart. It thrived best in rich, light soils, unmixed with clay and sand. Workers prepared the soil during the preceding winter by turning it with a plow or hoe and harrowing free of grass and roots. The cultivation was relatively simple, consisting of keeping the plants free of other growth. About four months after planting, when in full bloom, the plants were cut down for processing into dye, the average yield being slightly over forty pounds per acre.[27]

In Georgia the planting might be staggered so as to obtain three cuttings before the onset of frost at which time the harvest abruptly ended. The most disagreeable task was processing the stalks, for they emitted an offensive and nauseating odor. Because of the stench and flies associated with the industry, the Commons House of Assembly in 1774 required that the residual weeds be buried or destroyed after removal from the vats.[28]

The processing of indigo required two wooden vats, each approximately twelve feet square and four and a half feet deep. The harvested stalks were placed with water in the "steeper" vat where fermentation was completed in twelve to fifteen hours. The plants then were placed in a second vat called a "battery" where they were beaten and the water vigorously agitated for fifteen or twenty minutes until a curdle developed. At this point lime water was added to hasten the formation of the bluish precipitate. The beating, stirring, and agitation were continued until a strong purple color appeared; then the pulpy mass was allowed to settle for eight or ten hours. The water was then drawn gently from the battery so that the precipitate remained on the bottom. The latter was then strained through a sieve, put in osnaburg bags and further drained, and finally pressed out so as to be entirely free of water. The dried indigo was then taken out of the bags and cut into small pieces about two inches square and laid out for additional drying and airing in a house especially constructed for this purpose. In the drying-house the blocks were carefully turned three or four times daily to prevent rotting. During this period the dye might become infested with flies and insects and as a result be ruined.

Indigo had to be well dried before it was packed for market, in November or December. Merchants judged its quality by observing the closeness of the grain and the degree of brilliance in its violet-blue color. They might also place it in water to see if any solid parts were left undissolved. If the product sweated in the packing barrel it indicated insufficient drying and a high tendency to rot.[29]

The cultivation of hemp differed little from the cultivation of indigo and its processing was only a little less troublesome. The seed were sown broadcast on a moist acid-free soil and the plants grew to a height of four to ten feet. Reaped by hand, the stalks were soaked in soft water to free the fibers, which were then cleaned and plaited. A bounty to encourage the production of hemp was authorized in Georgia as early as 1762, and the crop was doubled in the following year. Six years later the bounty amounted to ten shillings per hundredweight of cured fiber and a smaller sum for dressed flax. While this crop never reached the importance of a leading staple in Georgia, it was among the significant crops exported from the province by the end of the Colonial period.[30]

Hemp and indigo were grown on land which also was suited to the non-staple crops. Growing these two staples not only cut down on the production of subsistence crops, but also tended to exhaust the soil after a few years of planting. Always interested in a balanced economy, the Trustees very early placed a bounty on food crops, including wheat, peas, potatoes, and corn, the last being the most important food crop grown.[31] This crop was particularly vulnerable to deer and roving cattle and, as was true with all other cultivated crops, the fields had to be fenced to a height of seven feet. After the ear was formed on the plant, there was the danger of nocturnal visits by wild rodents, so that planters often were forced to walk their dogs around their fields two or three times each night. In the earlier stages of the crop, similar safeguards were frequently required during the day against wild turkeys and crows. Harvesting corn entailed great labor for it had to be carried to the barn by hand, carting being impossible in newly cleared lands. Corn houses were built of skinned logs to lessen the danger of infestation by worms and insects. These airy structures were set on posts four feet above the ground.

The Colonial farmer planted corn in hills about six feet apart, allowing three or four plants to a hill. Between every two or three hills peas were planted, sometime in early June. When the corn ripened the farmer turned the ears downward by bending and partly breaking the stalk. This not only kept rain out of the exposed end of the ear, but it hastened drying and gave more air and light to the pea crop which now began to overspread the ground. Sweet potatoes or yams usually followed corn in successive years. These were planted on raised ridges, usually in July, from vines cut from older plants. After the crop was harvested, the potato field was used for grazing hogs.[32]

Sweet potatoes were cured in an oblong pit two feet deep over which was built a lean-to roof, with one end resting on the ground and the other on crutches. The pit was lined with corn stalks, on top of which the potatoes were placed and then covered with rice straw or additional corn stalks. The pit was then covered with earth to secure it from rain and frost, with an opening left in the top to provide ventilation.[33]

When food crops were in short supply the people complained. The failure of these crops during 1737-1741 was the source of most of the discontent associated with that period. In 1737, and again in the following year, drouth was particularly severe in the colony. At Darien in 1738 the expected corn crop of three thousand bushels did not exceed five hundred. At Savannah weather conditions were slightly more favorable, and the failure of the corn crop in that area was attributed to the planting of an unacclimated yellow variety imported from Pennsylvania. Toward the end of the summer heavy rains caused as much grief as did the earlier drouth.[34] In the early spring of 1739 the most severe freeze yet experienced in Georgia struck the coastal area, forming ice two inches thick on the ponds and damaging early vegetables.[35] As late as 1741 the inhabitants of Darien and Frederica still were dependent upon the Trust for food, and by the following year many of them had left the settlement. In that year the scarcity of corn and other food crops caused general complaints along the coast from Savannah to Darien. An invasion of caterpillar worms which attacked sprouting plants necessitated as many as four replantings, and William Stephens stated that corn could not be bought at this time at any price.[36] Savannah, with its less than a hundred and fifty houses and huts, was so reduced in spirit that one citizen complained that the town had "grown so thin of people, that few are to be met on the Streets . . . so that Planting went on heavily, with a small number such as were left at home."[37]

At this point Patrick Tailfer and David Douglas departed for Charleston and from there wrote a volume denouncing Trust officials and their policies, and Oglethorpe in particular. "[Either] the Drouth burns or Rain drowns the Corn," they wrote, "and makes the Peas fall out of the Pod; Deer, which no Fences can exclude, devour those little settlements in a night; Rats and Squirrels do the same; Birds eat the seed out of the Ground, and dig up the Blade after it is Spired; and Variety of Worms and Insects devour the half of it."[38] They attributed the low state of morale largely to inadequate food. With more redundancy than truth perhaps, they summarized the effects of a diet of salt meat combined with hard work under a torrid sun. "[It] created inflammatory Fevers of various kinds both continued and intermittent," they stated, "wasting and tormenting Fluxes, most excruciating Cholicks, and Dry-Belly-Achs; Tremora, Vertigoes, Palsies, and a long train of lingering nervous Distempers; which brought on to many a Cessation both from Work and Life."[39]

More optimistic reports came from the remote settlements in the upcountry. Augusta, founded as an outpost in 1735, appeared to be in a relatively thriving condition. Possessing forty families, a garrison of twenty soldiers, and three trading houses employing 500 horses in the Indian trade, the town carried on no agricultural experiments with exotic plants. Corn, small grain, and livestock were its principal agricultural crops.[40]

The German colonists at Ebenezer, situated on the extreme northern limits of the treaty line of 1733, were also in a thriving condition. Frugal, hardworking, and deeply loyal to their Lutheran faith, these former tinkerers from the mountains of Salzburg became the most successful farmers in Colonial Georgia. By 1738 all except three of the sixty-five lots at Ebenezer were occupied. Their diversified productions consisted of 1,104 bushels of corn, 429 bushels of Indian Peas, 518 bushels of potatoes, and 398 bushels of rough rice, in addition to pumpkins, cabbages, and other garden products.[41] During the following year, in a request to the Trustees to transport another group of their countrymen to Georgia, their life at Ebenezer was described as follows:

> Though it is here a hotter season than our native country is, yet

not as extremely hot, as we were told . . . but since we have been now used to the Country, we find it tolerable, and, for working People, very convenient; settling themselves to work early in the Morning, til Ten o'Clock; and in the afternoon, from Three to Sun-set; and having Business at Home, we do them in our Huts and Houses, in the Middle of the Day, til the greatest Heat is over. People in Germany are hindered by Frost and Snow in Winter, from doing any work in the Fields and Vineyards; but we have this Preference, to do the most and heaviest Work at such a time. . . . We were told by several People after our Arrival, that it proves quite impossible and dangerous for White People to plant and manufacture any Rice, being a Work only for Negroes, not for European People; but . . . we laugh at such talking . . . .[42]

In marked contrast to the British along the coast, not one of the Salzburgers had abandoned his settlement in 1740 and their thriving condition was a marvel to their English-speaking neighbors.[43] On a visit to Ebenezer in June the Rev. George Whitefield noted on a walk of four miles that he was "in almost one continual field," which was producing an abundance of crops for the subsistence of the people. He admired their singleness of purpose, "the strong helping the weak, & all seeming hearty for the Common good."[44] In 1742 the bounty paid on corn, peas, and potatoes in that part of the colony amounted to £273.[45] William Stephens wrote to Mr. Verelst, accountant for the Trustees, that "Ebenezer we see grow to such maturity as needs no further Leading Strings hereafter, & I wish it could be said of Savannah & its Neighborhood."[46]

Much of the success of the Salzburgers may be attributed to their spirit of enterprise, to the novelty of land ownership, and to their faith in the soil. With the exception of the silk industry, to which they applied themselves with remarkable zeal and a high degree of success, their agricultural experiments were confined largely to indigenous plants, including wild grapes, Indian peas, pumpkins, and maize, and they recognized the advantages of cultivating old Indian fields.[47] A Salzburg woman is credited with introducing rye, originally known as "German Corn," when she picked out three grains from Indian corn bought at Purysburg and planted them in her garden.[48] The pastor at Ebenezer, John Martin Bolzius, was as much an agricultural innovator as a religious leader. He was painstaking in his efforts to discover the best use for pine-barren land, on which he conducted an experimental farm. In 1745 he acquired a copy of Jethro Tull's *Horse Hoe Husbandry*, and at the close of his sermon each Sunday he instructed his congregation on the new process of cultivation with plows. He later requested the Trustees to send him Tull's wheat and turnip drill, since, wrote he, "No person here is able to build such an useful Engine."[49]

A marked characteristic of Colonial agriculture was the absence of plows and the scarcity of horses. "If the poor white people of this Colony," wrote Bolzius in 1745, "could be supply'd with a Horse each or 2 families, or a Couple of broken oxen, or with Convenient tools for Agriculture . . . they would not make complaints of the great Heat of this Climate."[50] An axe to clear the land and a hoe for cultivating it were the principal tools with which all farm work was done, the spade and the rake being secondary. It is to be hoped, wrote William Stephens in 1745, "[that] Plows will (in 2 or 3 years) be in fashion here, if not sooner, for Hoes are looked on as Badges of Slavery among the labouring English . . . ."[51] Stephens was unduly optimistic, however, for plows and other horse-drawn machinery were conspicuously rare in Colonial inventories, except on the large plantations, and even there they did not come into wide use until after the Revolution.

A typical inventory of a Georgia coastal farmer in the late Colonial period was that of Thomas Goodale who died in January 1762. He possessed a family of five slaves and three hundred acres, much of which was classed as good indigo land. His tools consisted of hoes, axes, hammers, and knives; a chisel, auger, whipsaw, cross-cut saw, hand saw, and a cleaver; four drawing knives, eighteen iron wedges, and a grinding stone. He had on hand 238 pounds of indigo, thirty bushels of corn, five horses, and thirty head of range cattle, a horsecart and a saddle, a still, 25,000 shingles, and three rawhides. Among his household equipment were silver shoe and knee buckles.[52]

This inventory suggests that much of Goodale's income was derived from forest products. Tar and pitch and turpentine (known as naval stores), processed boards, barrel staves, oars, shingles, and hoops were sizable exports from Georgia after 1750.[53] Less bulky and more valuable than heavy timber, the transportation of these commodities from inland farms was not a serious problem. Commerce in heavy timber was highly seasonal, coming to a virtual halt in dry seasons as a result of low water in the rivers.[54]

The importance of such auxiliary enterprises as the forests and vacant lands afforded is indicated in the list of exports from Georgia in 1763. In addition to 7,500 barrels of rice, 9,633 pounds of indigo, and 1,250 bushels of corn, the export list included deer and beaver skins, cowhides, naval stores, timber, and "other provisions" amounting to a total value of £27,021. Ten years later this total value had increased almost fivefold, and the list had expanded to include sago powder (a starch made from sweet potatoes), beeswax, tallow, hemp, tobacco, salt beef, and pork.[55]

The regulation of naval stores was a subject of Colonial legislation. The Commons House of Assembly in 1766 stipulated minimum weight and volume each of a barrel of pitch, tar, and turpentine, and also required inspection for purity and quality before shipment. Similar regulations later covered barrel staves and shingles, and cypress, pine, and oak timber.[56]

The basic techniques of the tar and turpentine industry have not changed greatly since Colonial times, except for improved methods of slashing the trees and better equipment for receiving and processing the gum. The bleeding of the trees began about the middle of March and continued until frost. The boxes were emptied of gum every two weeks, and one worker gathered two barrels a day from an average of 1,350 trees. The oil was obtained by a simple farm-yard distillery. Tar was extracted from pine wood by subjection to heat in an earthen oven built in the ground. Pitch, the solid part of tar, was separated from the viscous liquid by boiling.[57] Some potash was made from burned hardwoods, by a simple process requiring a wooden hopper in which the ashes were leeched with water and the lye drippings subjected to a high heat. Although England imported annually about 100,000 tons of potash, most of it came from Russia and Poland. The product made in Georgia was used locally in making soap and tanning leather.[58]

No consideration of the economic value to settlers of the vast public domain would be complete without including the early livestock industry. Cattle, all of which were grown under free-range conditions, were among the first purchases made by the Trustees, who acquired the original stock from South Carolina.[59] Several public herds were accumulated in various parts of the colony, the most notable one being located near Ebenezer, which numbered an estimated 700 head in 1741. A total of 2,000 were corralled in the spring of 1742, apparently collected from all Trust herds.[60] Since a range of twenty-five acres was considered adequate for one cow, it is likely that the total public range exceeded 50,000 acres.[61] In addition to the public herds, individual colonists possessed from less than a dozen animals to several hundred. These private herds were found throughout the colony, both on the mainland and on the islands, and apparently all had access to free grazing on the public domain. Colonists occasionally were advanced public money for the purchase of cattle to stock their lands.[62]

The Trustees' herd was the largest in the colony during the period, and its management entailed considerable responsibility on the part of officials. The employment of honest and experienced pindars (an early term for "cowboy") and providing them with horses was a constant problem. In 1742 the public cowpen keeper at Ebenezer, Joseph Barker, was convicted of fraud in connection with his official duties, but it was found impossible to replace him with one skilled in such work. As a consequence, he was merely fined and placed on probation.[63] His work being largely seasonal, the cowpen keeper was employed for £35 annually. He received an additional £37 for an assistant and for the maintenance of two horses, and £10 for the services of a farrier. He was charged with maintaining all private herds in the vicinity as well as those belonging to the Trust.[64] Occasionally pindars were employed on a fee basis, receiving five shillings for each cow driven up, a shilling for cutting one out of the herd, and a shilling for branding.[65] In 1739 the pindar at Savannah was furnished with six horses and charged to bring to the cowpen there "all such Heads as lay scattering many Miles wide, and not to be turned out again without being regularly branded and marked by the rightful Owners."[66]

The cowpen was an important institution in Georgia throughout the eighteenth and the early part of the nineteenth centuries. It was usually officered with a superintendent and a corps of sub-agents, all of whom were skilled horsemen, experienced woodsmen, and good shots with a rifle. Around the enclosure for stock a hamlet of cabins was erected, and in the vicinity was a considerable area of cleared land for the cultivation of corn. When the cattle were corralled, there was a noisy scene of activity in the midst of savage wilderness. These cowpen sites afterwards became important centers of settlement when farmers followed the cowdrivers in their search for cleared land.[67]

The annual roundup of cattle usually began around the first of March and continued until June. During this period six men and twelve horses might bring in a thousand head.[68] While the cattle were in pens undergoing branding or selection for market, it was a common practice to set fire to dry swamps filled with dead cane and other combustible matter in order to stimulate the growth of young herbage.[69]

Because of vast swamps and dense forests, generally not more than seventy-five per cent of the cattle could be brought in during the roundup season. Some might even become lost in the dense swamps and go unretrieved for several years. In 1738 a herd of one hundred wild cattle brought from Carolina broke out of the pen and were never recovered.[70] The animals were so wild that driving them was difficult, and it was often necessary to slaughter them on the range and transport the meat to market. In speaking of the Salzburger herd, Bolzius said: "They have many Heads of Cattle in the Woods, partly branded, partly not, & consequently wild.... Other wild cattle in the Neighborhood have carried away our tame Cattle from our Range, & our people were not able to bring them up without good Assistance of men, who understand the Woods, & are provided with strong Horses."[71]

The scarcity of horses was a constant problem for the range cattle industry. With only a limited number of horses being raised locally, it was necessary to buy most of them in South Carolina at high prices. An occasional roundup of wild horses was made from those having escaped from Indian traders or from the Spaniards in Florida. On one occasion horse hunters brought in five animals but complained that their own mounts had been ruined in the chase.[72] Georgia's supply of horses as well as cattle was augmented in 1740 when General Oglethorpe's troops captured and drove to Savannah several hundred head of Spanish cattle and a number of horses captured south of the Altamaha.[73]

Men skilled in riding and in judging horseflesh were as rare as good horses. When Trustee officials recommended Thomas Stephens to go to Purysburg in South Carolina to purchase mounts, his father remonstrated, writing that since his son had been in America only a year, he "feared to commit such a trust to him, at least he should be overreached; for allowing him to be a judge of the Make of a Horse and his goings, he might nevertheless not discover his hidden faults, and therein be Jockyed...."[74]

Other problems of Georgia's early range cattle industry were remarkably similar to those which prevailed on the western ranges more than a century later. The trans-Altamaha region, bordering on Spanish Florida, was a popular rendezvous for cattle thieves, cutthroats, and renegades. Many raids on Georgia herds originated in that quarter, even after Florida passed into British hands in 1763. Indians living close to the periphery of white settlements in the northern part of the colony also proved troublesome. "[We wish that] the Uchee Indians could be refrained from doing our Inhabitants Mischief," complained a Salzburger, "not only in their Gardens & Fields, but in their Hoggs, Horses, & Cattle likewise, which partly are shot by them for Meat & for the sake of Cow-&-Horse-Bells . . . ."[75]

Cattle stealing was by no means limited to Indians and border outlaws, however. The distress which accompanied the food shortage of 1738-1741 greatly increased this vice among indentured servants, who seemed to possess a peculiar craving for fresh beef.[76] Oglethorpe complained at that time of the vast numbers of the public herd that were being stolen and killed, and noted the general reluctance of juries to convict the offenders. He also complained of the runaway slaves, white servants, and outlaws from Carolina infesting the inland parts of the country. "[And] thieving for want of Rangers to pursue them is grown so Common," he said, "that great numbers of Hogs and not a few Cattle have been killed in the Woods, so that it is dangerous to let them out, and people have neither Inclosures nor food to keep them at home."[77] Such conditions caused him to extend for a considerable period the usual season for cattle hunting, by which "sixty wild Calves [were] taken up and marked at the Cow Pen at Ebenezer."[78] It was at this time also that the attorney general ruled that all unbranded cattle were to be considered the property of the Trustees.[79]

A more sophisticated form of stealing than outright poaching was that involved in over-branding and the branding of calves belonging to others. The public herd appeared to bear the greater burden of this type of cattle rustling. The Trustees' brand consisted of the letters G and C, separated by a small diamond. A horse bearing this brand, but showing that an attempt had been made to obliterate it, turned up in Purysburg. The rider, Gideon Mallet, was accused of placing G and M over the original mark. When charged with changing the brand, he claimed that the horse had been sold to him by an Indian. Despite the fact that such a transaction was unlawful both in Georgia and in South Carolina, the jury failed to indict him, on grounds of insufficient evidence.[80] On another occasion one Bradley, employed as a public rider, was accused of placing his own brand on Trust cattle, which "the letters G C would more rightly have been placed on." However, it was impossible also to convict him.[81]

More than 150 cattle brands were registered in the colony in 1755, and an average of nineteen was added to the list annually throughout the remainder of the Colonial era.[82] Georgia cattle brands of this period were not unlike those found on southwestern ranges in the nineteenth century. Generally they consisted of the owner's initials in letters which might vary in size and form, with an occasional addition of such markings as an arrow, a diamond, circle, square, or bar. John Gibbons marked his stock with a bow and arrow on the shoulder and the letter J on the buttock.[83] After a spirited argument between John Milledge and Jacob Matthews over the ownership of a steer branded J M the value of earmarks as a supplementary identification was demonstrated, a method of marking generally used only on hogs and sheep.[84]

Range cattle on the Georgia coast appear to have succumbed more frequently to disease than those of the upcountry. In the spring of 1741, during a dry season, a disease called "black water" appeared in epidemic form, its common symptoms being the discharge of urine of a deep red color. Some believed it was caused by feeding on a poisonous herb, but the malady disappeared with heavy rains which began in June.[85] The disease returned in the spring of 1742 and in 1743, again at a time when ponds were dry. In the latter year it was estimated that half of the public herd succumbed, most dying within two days from the onset of the disease. The epidemic returned in 1744, but the mortality rate was considerably reduced, and gradually the disease disappeared. The cause of the malady was never positively determined, but the herd at Ebenezer was generally free from its ravages, and little mortality was reported there.[86]

The open range not only invited epidemic diseases but also produced poor and scrawny cattle, since there was no opportunity for improving the herds by selective breeding. Among thirty-six branded cattle belonging to Richard Leake, for example, a wide range of colors is indicated, such as brown, red, black, brindled, red with white face, white with black head, and brown and white spotted.[87] The average weight of a cow apparently was under seven hundred pounds, and seldom did an ox weigh over eight hundred.[88]

In 1751, one year before the Trustees relinquished their charter, their cattle herds were sold to private individuals, the Salzburgers obtaining the largest number.[89] Some private herds grew to great size, and management was continued in much the same manner as under the Trust. Throughout the remainder of the century local legislation reflected the growing importance of the cattle industry. The exportation of pickled beef and salted pork, cowhides, and tallow was significant. Local tanning of leather was encouraged in 1768 when the assembly placed a tax on raw hides exported from the colony.[90] Eight years later attempts were made to prevent frauds in marketing beef and pork. Barrels in which they were packed were regulated as to quality and capacity, and those for beef should contain 200 pounds of well-cured meat exclusive of the head. Not more than one shank, one skin, and one-half neck could be placed in each container. Each barrel of pork, weighed and inspected by officials, could contain no more than two heads. Other laws had as their purpose the minimizing of cattle rustling. Reflecting such activity on the part of runaway slaves, Negroes were forbidden to drive, kill, or brand any horse or cow unless assisted by a white person.[91]

Of far less importance than cattle in Colonial Georgia were hogs, sheep, goats, and poultry. Hogs generally were not permitted within the limits of towns. Oglethorpe once ordered a number of these animals shot after they had wandered into Frederica from an outlying plantation and had damaged the fortifications by rooting into the earthen foundations.[92] Hogs were most popular when raised on some island where they could enjoy open range and yet be limited in their rovings. The presence of numerous "Hog Islands" on the local maps of the region today give ample testimony to the popularity of this practice.[93]

Goats appeared in 1741 when Robert Williams, a recent arrival, turned some loose in Savannah. They were led by an old he-goat who made holes in fences so that nothing was secure against him. Joseph Fitzwalter, whose garden was destroyed by these animals, killed the lead goat and proclaimed his deed to the populace. Later when Fitzwalter was driving home his geese, which also had free run of the town, the offended Williams killed one of his geese in revenge, thus provoking a war between goat man and goose man.[94] However, geese proved far more popular than goats in Colonial Georgia, for almost every settler came to possess them. In addition to providing a supply of feathers for winter mattresses, geese were thought to consume noxious weeds and grasses harmful to livestock and thus to improve the quality of grazing land.[95] The common varieties of barnyard poultry, so popular in a later era, were almost non-existent.

Sheep were introduced from Carolina in 1735 when a flock of forty was brought to Frederica. Thinking that they might be

COLONIAL AGRICULTURE   31

herded as English sheep were, they were turned out to graze, whereupon they reverted to wildness.[96]

The meat supply of the Georgia colonist could always be supplemented by wild life, of which there was a great variety. In addition to fish and waterfowl there was an abundance of turkeys, deer, and quail, and an incredible number of wild pigeons and other migratory fowls.[97] The buffalo might still be seen, but he was fast disappearing from the region, and did disappear entirely before the end of the Colonial period.[98]

III

The Passing of the Frontier

PRACTICALLY every male citizen in Georgia who had been loyal to the Whig cause during the Revolution qualified for a military bounty grant under one of the several laws passed during that conflict. Not only were native Georgians entitled to lands, but veterans from other states who had served in Georgia with the Continental army were also eligible. In addition, any Revolutionary soldier who was entitled to land under one of the bounty acts of the Continental Congress was extended the privilege of taking his grant in Georgia.[1] As a result, the state granted bounty land to a total of 9,000 veterans, of whom probably not more than 2,000 were natives of the state at the beginning of the Revolution.

The total acreage granted to all veterans during and immediately after the war, including headrights as well as bounty grants, probably exceeded 3,000,000 acres, of which 750,000 acres went to Georgia veterans. Grants of vast estates were made to a few high ranking officers and specially designated heroes, among whom were Elijah Clarke, Anthony Wayne, Nathanael Greene, and Count d'Estaing. The last, although a citizen of France, received a princely gift of 20,000 acres in Franklin County, made in four grants of 5,000 acres each. Georgia's benevolence in land was extended to veterans of the Indian wars and campaigns including those which occurred after the Revolution. Some of these land warrants were commuted to money payments.[2]

The settlement of veterans' claims was the most urgent business confronting Georgia officials following the war, and land distribution was the unifying theme of the state's politics for the next half century. The land law of 1784 increased the bounty grants to 287½ acres but removed an earlier tax exemption clause. Even though the state's title to the land was not yet confirmed, the counties of Washington and Franklin were opened for settlement in 1784, wherein most of the bonus lands were located.[3] Gently rolling to level in topography, these lands were highly prized, being ideally suited to corn, wheat, and tobacco. This area later became the earliest center in Georgia for the production of green-seed upland cotton. Except for the southern half of Washington County, the region was relatively free of pine barrens.

This area, lying between the Ogeechee and Oconee rivers, was the scene of feverish activity by those seeking land under the headrights law of 1784 and those seeking fulfillment of bounty claims. The dislocations of the war had resulted in the loss or destruction of many of the state's records. Many holdings had been abandoned and were now occupied by newcomers whose titles were unconfirmed. Common squatters and war refugees had also settled on unoccupied lands. These new settlers found themselves involved in heterogeneous and overlapping claims. Thus the land situation could hardly have been in a more chaotic condition. A land court office was established at Augusta for the purpose of handling the problems. Claimants had to obtain warrants from this office before lands could be located and surveyed, and the titles issued.[4]

Something resembling a "sooner movement" appeared at Augusta. An indication of the land fever prevalent at this time and the anxiety of the veterans to obtain the choice lands in the new counties is reflected in subsequent proceedings. In order to settle the question of priority in the issuing of warrants, the court devised a lottery system, the second of this type used in Georgia for the distribution of public land. The names of claimants were deposited in one box with an equal quantity of numbered paper tags in another. On the day appointed for the drawing of warrants, the names and the corresponding numbers were drawn from each box, which procedure determined the order of issuing warrants. In this manner all petitioners were placed on an equal footing in the selection and surveying of their land and in the recording of the title.

The great number of applicants overwhelmed the office and hindered the efforts of its officials in issuing the warrants. Becoming impatient with the system, a crowd of fifteen hundred to two thousand men assailed the office and threatened its keepers with physical violence. Breaking open the doors and swarming into the room, they seized the warrants irrespective either of number or alphabetical order.[5] The confusion resulted in the loss of several hundred warrants for which duplicates had to be made. Surveyors in the two counties were then ordered to execute surveys only on warrants whose proper owner could be certified.[6]

The law required that the plats be surveyed into squares and rectangles, the length of the latter not to be more than twice the breadth. The 287½-acre lots nearly all conformed to this description. The resulting land maps show the headrights or non-bounty grants to be exceedingly irregular. In both types of surveys the surveyors used blazed trees, meandering streams, stumps, and other perishable marks for boundaries and reference points, thus making it exceedingly difficult to establish accurate lines after the lapse of a few years. The maps also bore descriptive terms, such as "rice and indigo land," designations used in the older coastal counties whence the surveyors had come.[7]

One of the leading problems arising from this system was that of determining which land had already been surveyed and patented. Plots* often overlapped, or the land represented was so poorly surveyed and marked as to be difficult to locate. Also the confused patchwork of vacant lands lying between the appropriated areas made it impossible to delineate the public domain. The result was long-drawn-out litigation. Land courts, whose principal job it was to issue warrants, often found their time being consumed by hearing caveats. Also the administration of this system evoked clashes between large planters and small farmers.[8] The poor man who could not afford to make a personal exploration of the land to be selected for survey found that his warrant, or "headright," often was worthless.

When the surveyor's plot was completed it was filed in the surveyor general's office, and then the governor issued a title to the land. For this instrument, called "a plot and title," the applicant paid only the cost of surveying and of processing the papers. Those who were not entitled to bonus land and others whose warrants were for land in excess of the headrights allotment went through the same procedure, except that they were charged a nominal price for the land. The purpose of the headrights system

---

*The term "plot" is used herein where the reference is to the map or drawing. "Plat" is used to denote the physical area.

was the rapid settlement of the country rather than enrichment of the public treasury.

It was the advent of the large-scale land speculator and his corrupt influence upon governmental officials which brought about the end of the headrights system. The land speculator was attracted to Georgia after the Revolution by the state's vast public domain and by the vulnerability of its debt-ridden government to pecuniary enticements. This type of land seeker for the most part came from the older states where there was a more mature commercial life and where public lands no longer existed.[9]

Speculation first appeared when certain military officers began signing certificates of veterans who served under them in order to acquire the land for themselves. It was perhaps because of this abuse that warrants came to be limited to 1,000 acres for each individual under the land laws of 1783 and 1784. Fraudulent surveys were reported in 1784 involving land in Franklin and Washington counties with the result that some sentiment appeared favoring the centralized administration of the land laws.[10] However, the principle of local administration, favorable to the speculator, won ascendancy. The land law of 1789 removed the limitation of 1,000 acres on land grants, and soon afterwards the laws were amended in such a way as actually to encourage fraudulent surveys.

The authority to issue warrants and to act as a county land court was vested in three local justices of the peace. The warrants which they issued were directed to the surveyor of a particular county for running the lines and drawing up the plots. When completed, the plot was sent to the surveyor general for official recording and then to the governor who signed the grant and affixed to it the state's seal. After this procedure the plot and grant were returned to the owner as an official patent, or deed.[11] Thus the surveyor general was reduced to little more than a record clerk and, while the governor's signature gave the entire process an ostensible authenticity, the major responsibility for honest procedure rested with local officials who formed the county land court. Neither the governor nor his council had appellate jurisdiction over these courts, although the governor appointed its members.[12]

What is known as the Pine Barrens speculation began in 1790 when those who participated in it arranged to have themselves or their partners placed in office, either as county surveyors or as members of the county land court. During the next four years hundreds of grants were issued by them for millions of acres that simply did not exist. Most of these were issued in Montgomery County, which lay south of the fall line between the Oconee and the Ogeechee rivers, a barren pine wasteland with little prospect for future settlement. However, they were also issued in Washington, Franklin, Effingham, Liberty, and McIntosh counties.[13] The plots or maps indicated such reference points as oak, hickory, chestnut, and other hardwood trees which were not often seen in the Pine Barrens region, but which suggested to unsuspecting purchasers the presence of fertile clay soil. Other markings on the plot, such as streams, cleared Indian fields, springs and trails, also did not exist. The fraud was consummated when these plots and grants, bearing the governor's signature and the great seal of the state, were sold to buyers in Baltimore, Philadelphia, and other distant cities. As late as the twentieth century heirs to some of these bogus deeds still were trying without success to find the land described on the mapped plots attached to the original grants.[14]

The scope of these frauds is indicated by the fact that the total acreage "granted" in the counties of Montgomery, Washington, and Franklin exceeded seventeen and a half million, when the combined official acreage of these three counties at that time actually was less than a million and a half acres. In 1794 alone, 13,000 acres of bogus lands were granted by Governor George Mathews. Two of the largest grants, each for over a million acres, were made to county surveyors who pretended to have run the surveys for themselves or for each other. The records show that James Shorter managed to "survey" for himself 52,000 acres in one day and that, between August 6 and August 18, Thomas Cooper "surveyed" for Shorter 218,000 acres.[15] The Pine Barrens fraud had little political repercussions in Georgia, since no citizen of the state had been defrauded and no part of the public domain was actually transferred to outside speculators.

A transaction of an entirely different character was the Yazoo Fraud, which occurred in the same period and involved a public act of the state legislature. Its obliquity consisted, not in granting titles of ownership to the blue sky, but in the sale, through bribery and corruption, of millions of acres of excellent land in the far-western part of Georgia to a group of high-pressure speculators. Passing through a long gestation period, the act was climaxed in January 1795 during the last session of the legislature held at Augusta, before the capital was moved to Louisville. The state was heavily in debt, and the Yazoo sale was offered as a solution to this problem. What angered most Georgians, however, was the corruptibility of their elected representatives and the high-handed manner in which the speculators operated. The sale was made under an act whose title referred to such worthy objectives as the payment of state troops, the protection of the state's frontiers, "and for the purposes herein mentioned," yet its main purpose was to transfer to four land companies the title to more than 35,000,000 acres, or three-fifths of the entire western territory. For the sum of $500,000, which was one and a half cents an acre, some of the best land in the Old Southwest was alienated from the public domain. The tract began at the Mississippi River and ran eastward in a belt approximately 150 miles wide to the Coosa River, almost to the future site of Rome, Georgia[16]

The sale was negotiated in strict secrecy. It contained an ingenious bribe to Georgia citizens of 2,000,000 acres to be allotted them at the same price as that paid by the land companies. Some hard cash also passed as bribes into the hands of dispassionate legislators. One legislator obtained for his vote a promise of 75,000 acres of the land secured by the sale. Another sold out for a mere barrel of rice, while others gave their votes in exchange for Negro slaves. It is quite likely that the total amount of the bribes exceeded the purchase price of the land. Governor Mathews was reluctant to approve the bill but, after a few changes in the wording, it received his signature.[17]

The reform legislature which met at Louisville the following year, moved by an aroused public opinion, rescinded the Yazoo act with a *coup de grace* seldom employed by legislative bodies. Eventually, however, the anti-Yazooists found that it legally was impossible to invalidate the sale. In 1810 the United States Supreme Court in *Fletcher v. Peck* declared that the act of sale, being a contract, could not be voided by a succeeding legislature. Before the decision was rendered, however, Georgia had washed her hands of the business by ceding to the federal government all of her public lands west of the line of the lower Chattahoochee River, in which cession all of the Yazoo lands lay. The state received $1,250,000 for these vast lands, in addition to a commitment by the federal government that it would secure from the Creek and Cherokee Indians for the use of Georgia citizens all the land in Georgia east of the line of the lower Chattahoochee.[18]

This agreement was made in 1802. It was more than a quarter of a century, however, before the last of the Indians were removed. The most significant result of the Yazoo controversy was to bring to an end the old system of land distribution which virtually had been in effect since the beginning of royal control, in 1752.

More than thirty-one million acres of land still remained in the public domain after 1802. How to organize this territory and transfer titles to settlers was the basic problem which confronted the legislature as it met in special session the following year. Josiah Tattnall, the out-going governor, strongly opposed a policy which would enable a few individuals to monopolize land.[19] In true Jeffersonian style he argued the need for a large yeoman population to increase the militia potential, "the chief reliance for liberty in a democratic state," and he added that this would also guarantee the supremacy of whites over the blacks.[20] Tattnall's successor, Governor John Milledge, pursued a similar argument and in addition expressed his purpose of "securing the industrious poor against the grasping hand of the speculator." Both of these men favored free allotments in farm-sized parcels.[21]

Others believed that the land should be sold to settlers in limited amounts and at reasonable prices. Many of these were genuine liberals who would use land revenue to promote education and to develop the state's latent resources. Large property owners believed that land revenue would enable the state to eliminate taxes and would also secure a sound fiscal system. Some of these men held state revenue certificates and depreciated state bonds and not a few were also land speculators. Veterans who sought additional bonuses in land were also clamoring for consideration.[22] Thus it was not a simple problem to resolve these differences.

The new land law, passed in 1803, represented the triumph of a liberal policy of free land and equal grants in farm-sized lots. Only a recording fee of four dollars per hundred acres was to be charged. The land was to be surveyed and organized into counties before it could be granted. The problem of distributing the land equitably to the state's 162,000 inhabitants was solved by the old device of a lottery, now employed for the third time in the distribution of the public domain. Every free white male inhabitant of the state who had reached the age of twenty-one years and who was a citizen of the United States and a resident of Georgia was entitled to one draw. If such a person had a wife or a child he was entitled to two draws, as were widows with minor children. Minor orphans or a family of minor orphans were also entitled to a draw.[23]

The identification number of each lot was written on a piece of paper and placed in the hollow rim of a wheel set up to rotate in a fashion to cause the numbers to change positions. The names of individuals or families eligible to draw were placed in a separate wheel, designed in a similar manner. The name of the drawee and his corresponding lot were taken from separate wheels, thus to determine ownership of the lots. As formerly, the governor made the grant of title.[24]

This method, known as the lottery system, was used to distribute what remained of the state's public domain after 1802. There was a total of seven lotteries, beginning in 1805 and extending to 1832, from which more than 100,000 fortunate individuals or families benefited.[25] The vast area encompassed by the lotteries would have sold for more than $40,000,000 at the prevailing minimum price which the federal government charged for its public land during this period.[26]

No basic changes were made in the lottery system after 1803. The size of the lots varied from 40 to 490 acres, depending upon the ostensible value of the land. A considerable mountainous area in the northern part of the state, formerly occupied by the Cherokee Nation, had some gold-bearing possibilities, and it was surveyed into forty-acre lots, while in the remainder of that area each lot contained 160 acres. In Hall and in part of Habersham, Walton, Gwinnett, and Early counties the lots were 250 acres. Throughout southern Georgia east of Early County, in the original counties of Irwin, Appling, and Wayne, and in part of Habersham and Rabun in the northeastern mountain area the lots were 490 acres. In the broad Cotton Belt, which traversed the middle of the state and which included almost half of the entire region, the English system of surveying squares of forty-five chains was followed, producing lots of $202\frac{1}{2}$ acres.[27]

Successful drawees, except for Revolutionary War veterans, were excluded from participation in succeeding lotteries. Disabled veterans and war widows were sometimes given extra draws, and in time the eligibility list was extended to include children of convicts, idiots, the deaf, dumb, and blind, and also illegitimate children. The administration fee increased steadily, becoming fixed in 1820 at eighteen dollars per grant. Fractional lots were not placed in the lottery, but sold directly by the state, as were "reverted lots" and lots proved to have been drawn fraudulently. The state realized a net return of less than seven cents an acre on all the land it distributed under the lottery system.[28]

This new land policy represented a sharp return to the radical liberalism of the Revolution wherein it was claimed that all the people should have free and equal access to nature's productive powers. It represented the Jeffersonian ideal of a citizen's receiving the full product of his own labor and owning enough land to employ his industry and talents to the fullest extent. Unlike the Colonial land system, no special provision was made for planters who might need an extensive holding for the employment of a large number of slaves. However, such citizens found it easy to purchase extra land at reasonable prices from those who had been successful in the lotteries.

The most dramatic event in Georgia history in the thirty years following 1802 was the removal of the Indians from the western and northern portions of the state and the rapid settlement of the more desirable lands in these areas. The Creek territory extended from the Florida border northward to an uncertain latitude approximately at the future site of Atlanta. Constituting a vast acreage, this area embraced some of the finest cotton lands in the state. The presence of the Creeks within the chartered limits of Georgia was a source of general complaint among land-hungry citizens. While not so civilized as the Cherokees in North Georgia, the Creeks were cultivating most of the white man's crops, including corn, tobacco, beans, peaches, melons, grapes, and strawberries. They kept large herds of cattle as well as horses, hogs, and poultry flocks. The loom, the spinning wheel, and the anvil were found in some of their establishments as were a wide variety of implements for the cultivation of the soil.[29] Many of their tribal leaders were of mixed blood and a few owned large numbers of Negro slaves.[30]

Because the Georgia Indians were fast acquiring the political and cultural skills of their white neighbors, they promised to become assimilated, biologically and culturally, into the Anglo-Saxon pattern. Federal policy since 1790 had appeared to Georgians as lending encouragement to this result, which they violently opposed largely because of its implications in the final distribution of the valuable lands.[31] Between 1802 and 1806 three Creek cessions and two lotteries moved the boundary of the Georgia settlements from the Oconee westward to the line of the Ocmulgee River and westward from the coast to include the Talissee* strip, originally organized as Wayne County. By 1821

---

*This word is spelled variously as *Talise, Talisee,* and *Tallassee.*

other cessions and two additional lotteries embraced all of southern Georgia and placed the boundary of the middle belt of Georgia at the Flint River. After serious altercations with federal authority, Georgia was rewarded by a treaty with the Creeks in 1826, by which they ceded their remaining lands west of the Flint River with the exception of a small trans-Chattahoochee area in the northwestern portion of the Creek territory, which they ceded in the following year.

The fertility of this region attracted a large number of slaveholders from the older counties and the Chattahoochee and Flint rivers gave it a prime trade outlet.[32] Columbus, springing up almost overnight at the head of navigation on the Chattahoochee, became the mart for a wide and extensive area, reaching half the distance to Macon on the east and as far as Habersham County on the north.[33] The wide belt across the middle of the state, with the fall line crossing its southern portion, was known as Middle Georgia. Beginning at the foothills of Wilkes County and extending southward just below Augusta, it swept westward to the cretaceous area about Columbus. Its soils were largely red clay and gray sand with considerable potash.[34]

The northwestern portion of the Creek cession, lying in a triangle between the Chattahoochee and the Alabama line, was organized originally as Carroll County. A region of rolling foothills, this area, unlike Middle Georgia, was settled extensively by small farmers, who developed a subsistence economy. Census figures indicate that, while they originated largely in the older counties of the Piedmont, their mode of living conformed closely to the pioneer pattern. They were much below the average age of other Georgians and there was a disproportionately large number of males among them. Their agricultural productions did not show a strong allegiance to cotton until a decade before the Civil War.[35]

By 1820 Georgia's population had increased to 340,987, of which 43 per cent were Negroes. The flow of population into the Creek lands after this date was perhaps without precedent in the history of the state. Within three years after the lottery of 1827 the counties of Fayette, Coweta, Troup, Talbot, and Harris in the northern part of the 1826 Creek cession, each had a population exceeding 5,000 people.[36] Ten years later these counties contained a total of 133,000 inhabitants. The people came largely from the older counties of Middle Georgia and they brought with them large numbers of slaves and the agricultural patterns of the older communities. At the same time, there was considerable in-migration of people, notably from Virginia and the Carolinas. Some of these found homes on the older lands of Middle Georgia abandoned by their former owners for the newer lands of western Georgia and Alabama.[37] In the area around Milledgeville, Sandersville, and Sparta were vacant farm houses on cleared and fenced lands which passing emigrants could purchase at attractive prices. In 1828, for example, a farm of 150 acres in Hancock County was advertised for two years' accumulated taxes amounting to $1.32. A fifty-acre tract on the Little Ogeechee could be acquired by taking up a *fi fa* of fifty-three cents, and a thirty-two-acre tract on Shoulderbone Creek was available for a total price of sixty cents.[38]

Mrs. Anne Royall, visiting Milledgeville in 1830, observed that the capital of Georgia was "entirely deserted by men of business" and doomed to subsequent decay because of the rapid settlement of the fertile lands beyond the Ocmulgee. She expressed surprise that people from the older states did not come in even greater numbers to settle on lands in western Georgia.[39] In contrast to the drowsy appearance of Milledgeville, Columbus, according to the English traveler George W. Featherstonehaugh, was reeling under the boom of frontier activity, its streets swarming with drunken Indians and with prostitutes of both races.[40] Food prices were high. Corn sold for $2.50 a bushel, more than four times the price it was bringing in the older communities east of the Ocmulgee.[41]

Despite the booming activity in western Georgia, land was the cheapest and most plentiful commodity to be found there. Records of land sales in the new western counties in 1830 show that its price frequently was below ten cents an acre. A rifle, a dog, or a sheep sometimes exchanged for an entire lot of more than two hundred acres. A bushel of corn meal once exchanged for twenty-five acres,[42] and an innkeeper at Carrollton rejected a 202½-acre lot in payment for a night's lodging proffered by one who arrived the previous day to view the land he had drawn in the 1827 lottery. Land had become so plentiful by 1830 that only two-thirds of the grants made under the lottery of 1820 had been claimed. As late as 1854 unclaimed grants were being sold at sheriff's sales throughout many parts of southern and western Georgia. Despite these facts, Georgia's population between 1800 and 1830 increased more than that of any other state on the Atlantic coast from Maryland to the territory of Florida.[43]

Unlike Middle Georgia, much of the southern part of the state was covered with pine forests which had come to be identified with poor soil, and this part of the Creek cession of 1814 failed to attract its share of immigrating planters. Lying in the southeastern part of the state, beginning where the Cotton Belt ends, is a vast region of pine barrens and wiregrass, about the center of which is the confluence of the Oconee and the Altamaha rivers. This region's agricultural development also does not articulate with the story of Georgia's expanding cotton industry in the period before 1860. Until extensive use of commercial fertilizers made it more desirable to farmers, the most important crop of the region was yellow pine, tar, and turpentine.[44]

The people of southeastern Georgia were described by early travelers as ignorant and crude, and possessing characteristics revealing to later observers the symptoms of hookworm, malaria, and pellagra.[45] Emily Burke described a family in this region as living in an uncomfortable log house equipped with a bed and a table, but without chairs and possessing only the minimum of cooking utensils. Their food consisted of black coffee, pork, and corn bread made of meal and water. She saw no vegetables, butter, or milk. She observed that their meat was salted and dried and generally tainted, and she found it difficult to sit at the table where it was served. She thought the people of this region were a century behind their times in social and economic advancement.[46] As late as 1845 an Irish laborer described the natives of Irwin County as living in a wilderness after the manner of early frontiersmen. They subsisted principally upon wild cattle and hogs. They had no roads and few conveyances. Their sluggish streams did not operate water wheels, and they cracked their corn in Indian fashion between rocks, or else carried it forty miles to be ground into meal. "They say little, despise to be encroached upon by settlement, live on their flocks, and . . . exhibit many of the habits of the savage," he wrote.[47] On the eve of the Civil War no single town of any consequence had grown up on the Central Road between Macon and Savannah. Most of the stations had numbers instead of names, and they were devoid of houses or other signs of habitation.[48]

In 1830 the Cherokees still occupied northern Georgia except for an area in the extreme northeastern corner of the state which had been ceded in earlier treaties. The Cherokee country was mountainous and rugged, except for its numerous fertile valleys, with soils composed of disintegrated limestone. These lands did not greatly attract cotton planters and slave-holders, but it was a prime area for the production of grains and livestock. The Cherokee population was estimated to be 15,000 in 1820, half of whom were said to be of mixed blood. Some of them owned slaves, lived in comfortable houses, and their property in livestock and farm equipment was valued at more than half a million dollars.[49] In 1828 Georgia extended her jurisdiction over these people and took steps to extinguish their claims to the lands which they occupied. About this time the discovery of gold at Dahlonega provided an added incentive for their quick removal. In 1832, three years before the treaty of cession was actually consummated, Georgia surveyors plotted the entire area, and in the following year it was distributed in the last of the state's land lotteries. While the cession was not confirmed until 1835 and the departure of the Cherokees was not completed until three years later, Georgians immediately swarmed into the region to possess the land which they had drawn in the lottery and to acquire the better farms on which the Cherokees had made valuable improvements.[50]

The Cherokee lands of North Georgia were settled only a little less rapidly than were those of the Creeks. In 1840, just two years after the final departure of the last of Georgia's Indians, there were 62,878 people in the thirteen counties created out of the 1835 cession. Approximately half of these lived in Cobb, Cherokee, Cass, and Walker counties, where the best valley lands were found.[51] Rome, founded in 1834 at the head of the Coosa River, stood at the southern gateway of this western valley region. The town became an important connecting link between Middle Georgia and "Old Tennessee." It served as a market and a clearing house for corn, wheat, and livestock from the north, and cotton from the Coosa Valley, which begins just west of the city. "Cherokee Georgia," as it was called, contained no large cotton plantations except in its southern extremities, and only a few slaves.

The agricultural chronicles of Georgians who lived in the mountain coves in the northern part of the state may be related apart from the general story of agriculture in the more fertile valleys of that region. Isolated and crude, and with inadequate transportation, they remained ignorant of the outside world. An inhabitant of the region described its early settlers as a "rude, untamed and restless people" who moved thither from the foothills of Georgia and the Carolinas. They built their cabins, grew small quantities of corn and vegetables, raised scrub cattle, and with their rifles kept the family in meat.[52]

Among the early settlers of Forsyth County in the 1830's were the parents of Hiram P. Bell. He described his neighbors, living on the lower edge of the Cherokee country, as plain in dress and manners. The cotton patch and flocks of sheep kept the family in homespun clothing, made possible by deft hands in skillful use of the spinning wheel and loom. In the course of time a herd of livestock, an orchard, garden, and dairy added variety to their fare, and they lived well from a life of honest toil. Bell described the vicissitudes of pioneer life in that region beset with rattlesnakes. Capricious weather often caused crop failure and brought dire privation to comparatively isolated settlements. As late as 1846 Bell's father hauled cotton in an ox cart from Cumming to the terminus of the Georgia Railroad at Madison and sold it unginned for two and a half cents a pound. One year he planted a ten-acre tobacco crop, but it failed for lack of a favorable climate and because of his inexperience, whereupon he returned to cotton to supply the little cash needed for family requirements. A crop of wheat was threshed laboriously at the rate of five bushels a day.[53]

The eastern part of Georgia's mountain area did not figure greatly in the agricultural development of the state before 1860. The region, with a soil base of granite and limestone, was centered around Clarksville, the seat of Habersham County. This village early became a resort for tidewater rice planters, escaping the summer miasma of the Georgia and South Carolina coastal areas.[54] Nacoochee Valley, which lay in the vicinity, took on some of the characteristics of a Middle Georgia plantation community in the two decades before the Civil War.

Charles Lanman visited northeast Georgia in 1848 and wrote a vivid description of its beautiful scenery, interspersed with clear streams, waterfalls, and small strips of fertile valleys. Only at Tesnatee Gap in Habersham County did he find a single farm of such size and condition as to justify the description of a plantation. The owner, Francis Logan, master of twenty-four slaves, had acquired his wealth in the goldfields at Dahlonega, and he complained of the poverty and ignorance of his neighbors. Lanman found the mountaineers living in relative comfort, however, even though they had large families and occupied small, crude cabins. The principal productions of their land consisted of corn, wheat, rye, and potatoes. Their food supply was supplemented by an abundance of wild game and fish.

Lanman spent the night near Tallulah Falls with Adam P. Vandiver, a colorful mountaineer, whose log cabin nestled in the center of a beautiful valley through which the Tallulah River flowed. Vandiver was described as about sixty years old, small in stature, crude, illiterate, and the reputed father of thirty children. During the winter he engaged in hunting and trapping. In the summer he tilled with his own hands the few acres of land around his cabin. At the time of the visit Lanman found Vandiver's livestock to consist of "a mule and some half dozen goats, together with a number of dogs."[55]

Vandiver apparently prospered during the following decade. In 1860 his land was valued at more than eighteen hundred dollars. His livestock included horses, mules, cows, sheep, hogs, and goats. His farm produced wheat, rye, oats, corn, peas, barley, orchard and garden products, tobacco, honey, milk, butter, and fifty dollars' worth of home manufacturing.[56] When Vandiver died in 1877 he left an estate which belied the estimate of him, made by his Yankee guest thirty years earlier, and he might well have commanded the respect of his aristocratic neighbors from the lowlands, lounging in their summer homes at Clarksville.[57]

By 1830 the state's center of gravity in population, wealth, and political influence had shifted to Middle Georgia. Milledgeville, standing in the heart of this region, had been the state capital since 1807 and within a radius of ninety miles of this town lived two-thirds of the state's people. This area was fully committed to the growing of cotton, and navigation below the fall line had been opened on all of its principal streams.[58] Many people of foreign and northern birth were settling in the river ports along the fall line, notably at Augusta, Milledgeville, Macon, and Columbus.

The Cotton Belt in the middle portion of Georgia with its counterpart in southwestern Georgia and in the lush valleys of the former Cherokee country provides the main setting for the story of Georgia agriculture after 1830. This region was blessed with soil, climate, and topographical features which early brought into notice its agricultural possibilities. It became the home of a commercially-minded and cultured people. Its early trade centers gave its rich cotton lands easy access to the markets of the world and made farming a capitalistic enterprise. Here lived the great political figures which Georgia produced in the late ante-bellum period, and here also lived the men who figured largely in the state's annals of war, commerce, letters, education, religion, and agriculture.

IV

The Rise of Upland Cotton

THE AMERICAN REVOLUTION was followed by the withdrawal of British subsidies to growers of hemp, indigo, and rice. Although steadily declining in importance, these crops continued to be grown in Georgia throughout the remainder of the century and well into the next. Hemp, third in rank among Colonial staples, was the first to disappear when Yankee clippers began bringing cheap fibers from the Orient to the markets of the world. Small quantities of indigo continued to be grown until after the Civil War, although competition from British growers in the East Indies had destroyed its commercial position in Georgia by 1800. The rice industry held on tenaciously for more than a century after the Revolution.[1]

Rice was the slowest to recover from the ravages of war because of the loss of slaves carried off by British loyalists and also because of the physical destruction to plantations and equipment. A severe freshet in 1784 and again in the following year seriously limited the crop and delayed further its recovery from wartime disruption.[2] By the end of the eighteenth century it had recovered its pre-war position, but another series of floods and hurricanes in the first quarter of the following century contributed greatly to the hazards of growing rice on the Georgia coast. The first occurred in September 1804, when the most severe storm since 1752 hit the Georgia sea islands and devastated much of the mainland. Between 1804 and 1825 occurred eight or nine gales of sufficient intensity to do serious damage to crops.[3] In this period planters came to recognize the relationship between flooded rice fields and outbreaks of malaria and yellow fever. Fear of the two diseases caused the production of rice to be abandoned near centers of population and encouraged planters to leave their plantations in the hands of overseers during most of the growing season.[4] Despite these factors, Georgia's rice crop stood second to South Carolina's throughout the period preceding the Civil War.

Rice growing in Georgia reached its peak in the 1850's when approximately 30,000 acres were devoted to the crop. The production increased from 13,000,000 pounds in 1840 to 52,000,000 pounds in 1859, and the industry represented the largest type of plantation enterprise in the state. While rice Negroes were cheaper than those on cotton plantations, one prime hand was required for seven acres of rice, and rice lands in the 1850's cost eighty dollars an acre. When machinery and other necessities were added, the grower might have upwards of $100,000 invested. Many small planters cultivated a few acres of rice along with cotton, corn, and potatoes.[5] Bearded rice (so named because of its long and pointed husks), which could be grown under dry culture, was introduced into Georgia in 1809 when Thomas Jefferson sent to John Milledge some seed which he acquired in France.[6] Known also as upland rice, small quantities of it were grown on cotton plantations throughout the Piedmont in the period before the Civil War.

The rice industry in Georgia never completely recovered from the damage it received during the Civil War when Union troops occupied the coastal area and wrought injury to plantations and equipment. A great hurricane in the early part of the twentieth century, which rendered the fields sterile with salt, finally brought to an end Georgia's oldest staple industry.[7]

The coastal area rapidly lost its economic as well as its political importance after the Revolution, when the population center shifted to the upper reaches of the Savannah River with Augusta as its new focus. In 1790 the first official census returned a total population of almost 52,000 people in the five upcountry counties of Burke, Richmond, Wilkes, Franklin, and Greene. Less than twenty-two per cent of these were Negroes. In the six counties along the coast, including Effingham, there were only 31,083 inhabitants, of whom more than fifty-five per cent were Negro.[8]

Few settlements ever developed in the area south of the Altamaha, and as late as 1806 the road southward from Savannah was one of the poorest in Georgia. The stage went no farther than

Darien, from where mail and passengers traveled southward by sea. Great numbers of livestock still roamed the area, many individuals owning herds of more than a thousand head.[9] John Melish, at the beginning of the new century, observed the area around Savannah to be sandy and barren and contributing little to the needs of the people. Traveling northward through the Savannah-Ogeechee corridor he found the country improving rapidly. Near Waynesboro the road became elevated and dry. Pine trees changed to oak and hickory with a thick undergrowth of bushes and shrubs, indicating to him a high degree of fertility. In Burke County he found flourishing peach orchards and observed the first water-powered mills for grinding grain. Farther in the upcountry he found fields of wheat and corn, and so many settlements that he was never out of sight of a farmhouse.[10]

New crops developing after the Revolution were cotton, tobacco, and sugar cane, in order of importance. Sugar cane appeared on the coast at the beginning of the nineteenth century, when it was planted along with other crops, and the crop later was extended into the interior for more than a hundred miles. The total number of sugar plantations in Georgia probably never exceeded two hundred, and sugar cane failed to reach great commercial importance. The operational units were small, much of the product being grown for home consumption. Typical of the larger operations were those of Major Butler, who cultivated eighty-five acres, and John McQueen, who planted slightly less than fifty. The former employed seventeen hands and produced 140,000 pounds of sugar and seventy-five hogsheads of molasses.[11] While Georgia stood fourth in rank among the states in sugar production in 1860, her annual production was only approximately one thousand hogsheads.[12]

Tobacco for home use was grown widely during the Colonial period and some was shipped to Pennsylvania in exchange for iron, flour, and flour products such as keg biscuits (crackers) and ship's bread.[13] By 1785, Georgia tobacco had reached the European market in relatively large quantities. It was at this time that some attempts were made in Georgia at commercial processing, such as pounding into snuff, cutting for smoking, and plaiting into pigtails for chewing.[14]

The post-Revolution increase in the growing of tobacco was the result of the rapid settlement of the upcountry, where it was the first principal money crop. Farm inventories reveal that it was grown on nearly every farm in the area north and west of Augusta during the latter quarter of the eighteenth century and into the early 1800's, when the popularity of upland cotton caused tobacco growing to decline.[15]

By 1800 tobacco markets and warehouses for the public inspection of tobacco were located throughout the upcountry. Augusta had two such warehouses, while at Harrisburg, two miles north of that city, was a third. Petersburg, situated on a point of land where the Broad River entered the Savannah above Augusta, contained two tobacco warehouses. Edenborough and Lincoln, farther north, had one each. There were also one at Louisville on the Ogeechee, one at Montpelier, and one at Federaltown on the east bank of the Oconee.[16] Farmers brought their tobacco to market in hogsheads, rolling them along roads which followed the higher ridges to avoid moisture damage in crossing swamps and streams. These "tobacco roads" converged at some point on a navigable stream, usually at Augusta, where the product could be transported by water.

While most of the tobacco produced for market was grown within a few miles of a navigable stream, nearly every farmer in the up-country planted small quantities of this crop. The inventory of Alex Steed, a frontier merchant and jobber in Greene County during this period, indicates that tobacco was one of many items which the upcountry farmers exchanged for the few necessities procured from merchants. Skins of deer and beaver, and small furs, constituted the largest item of bartered goods, amounting to a total value of $260.00. Forty-five cow hides were valued at $67, and tobacco, of which there were 786 pounds, was valued at $16.[17] The farmers grew large quantities of corn, which they consumed at home or sold to the militia for seven to eight shillings per barrel. As much as seventy bushels of wheat might be grown on an acre of virgin land, and soon Augusta became Georgia's first milling center, thus diminishing the demand for flour products from Pennsylvania and Virginia.[18]

A wide variety of livestock grazed the open ranges of the rolling, wooded upcountry. At the end of the century carts and iron plowshares were common, as were horses and oxen. Cattle, hogs, sheep, beehives, and geese appeared on almost all farms. Cotton gins and "gin saws" were coming into use, but the flaxwheel was disappearing. Many farmers possessed a single family of Negroes and a few had acquired from twenty to fifty, denoting the rise of a class of upcountry plantation masters. Typical of the latter was Zephemiah Beal of Richmond County, who, in 1801, owned thirty-one slaves, as well as thirty-eight cows, a yoke of oxen, six horses, 241 hogs, and two sheep. In addition to saddles, he owned a riding chair and harness, a tobacco flat of seventy hogsheads' burden, and a branding iron. Stored in his barn as late as May were 1,600 pounds of corn and twenty-five bags of cotton, each weighing approximately 225 pounds.[19] A more affluent planter was Joel Early of Greene County. While he possessed only $4,240 in cash, he owned forty-nine slaves and held Georgia bounty warrants calling for more than 21,000 acres. In addition he held script in three Yazoo land companies totaling 259,000 acres. More of a land speculator than a planter, his agricultural productions, while they included no cotton, nevertheless embraced the widest variety of food crops, livestock, and poultry.[20]

The most significant feature of Georgia agriculture in the half century following the Revolution was the development of cotton as the principal staple crop. Experiments in growing this staple were made during the Trustee period. "I thought Cotton deserved a Place not too scanty" wrote William Stephens in 1740; "at leastwise I would try, whether it would turn to any Account or not."[21] Perennial tree-cotton from the West Indies was first planted, but it tended to die during the winter and required four years to bear fruit.[22] The annual plant was then introduced, and it produced such a quantity of seed that the cleaning, which had to be done by hand, was a tedious and laborious process, a work described as suitable "only for decrepid [sic], old People and little Children."[23] The difficulty of hand cleaning prevented this fiber from achieving commercial importance in the Colonial period.

What came to be known as sea island cotton, called "seabuird woo" by the Scotch, was first grown on St. Simons, "probably in 1778," according to Thomas Spalding of Sapelo Island. However, in the winter of 1785 or 1786 parcels of seed from a superior type of plant were sent to Georgia from the Bahamas, and soon afterwards production increased. Grown initially on St. Simons and Cumberland islands, sea island cotton quickly spread to other islands on the coast.[24] By 1790 substantial shipments of this cotton were being made, and planters, suffering from declining indigo prices,[25] found the crop an economic boon.

The cotton originally was planted on flat land, the plants standing five feet apart, and all cultivation was done with a hoe. One hand could work four acres, which might produce 400 pounds of lint. Later the plants were increased eight-fold in number and 350 pounds of lint to the acre were harvested. The crop generally was worked four times, and fields were cultivated in alternate years. By 1800 the practice of planting in drilled rows made on the crest of ridges thrown up by plows was generally followed. The rows were five feet apart and the plants stood from six inches to more than a foot apart in the row, depending on the fertility of the soil. In August the plants were topped to encourage better fruiting.[26]

The seed of sea island cotton were smooth and black and they could be separated from the fiber by forcing between two wooden rollers, by which process 300 pounds of lint could be cleaned in one day. The lint was then packed by a hand pestle into round bags, each containing approximately 300 pounds and occupying a volume of four and a half square yards.[27]

The type of cotton known on the coast as "green seed" possessed the ability to thrive under a wide variety of conditions, including those which existed in the Georgia upcountry. Because of the tenacity with which the lint clung to the seed it was impossible to separate the two by any means yet devised except laborious hand cleaning. As a result only small quantities were grown and they were devoted largely to home use.[28] To separate a pound of this cotton from the seed required a day's work. The task usually was done during winter evenings, in a family circle around the fireside. The commercial importance of green-seed cotton did not begin until Eli Whitney devised an effective machine for separating the seed from the fiber. Prior to Whitney's invention the only mechanical device used for this purpose was a simple bow string around which the lint was wrapped and then pulled until it snapped. Hence Georgia upland cotton of the early period was known as "bowed cotton."[29]

Whitney invented his "gin,"* while living at the Nathanael Greene plantation near Savannah. The device underwent a successful test in the early spring of 1793, when, in its initial stage, it proved capable of cleaning fifty pounds of lint in one day.[30] The gin consisted essentially of a cylinder about two feet long and six inches in diameter into which circular rows of small iron spikes were driven. As the cylinder was turned by hand, spikes reached through narrow slits into a drum filled with seed cotton, drawing out the fiber and leaving the seed inside. Various improvements

---

*The origin of the term is uncertain. Some early writers have stated that it is a contraction of the word "jenny" from James Hargreaves spinning jenny invented in 1765. Others claim that it is a shortened form of the word "engine."

were added later, such as brushes to remove the fiber from the spikes, and fans to clean the brushes, and more efficient gearings, but the basic principle remained unchanged. In 1795 the "saw gin" appeared as a rival to Whitney's device, although it differed from the latter mainly in having teeth that were cut into circular rims of iron instead of being made of spikes.

By 1796 Whitney and his partner, Phineas Miller, had thirty gins operating in eight communities in Georgia, some powered by water and others by horses and oxen. The two men had entered into a partnership with the intention of monopolizing the ginning business. Thus their profits would be derived from tolls charged on cotton processed rather than from the manufacture of the machines or the sale of the patent rights. The fact that the machine was so much in demand and so easily constructed, once the basic principle was understood, deprived Whitney of any substantial remuneration for his inventive genius. Gins of surreptitious manufacture were erected in every part of Georgia and South Carolina, and juries were unwilling to convict the manufacturer for infringement on the patent. More than sixty suits had been filed in Georgia before a single decision on the merits of Whitney's claim of infringement was obtained. This came in 1807 when a United States District Court at Savannah granted a perpetual injunction against one Arthur Fort. At this time more than thirteen years of the Whitney patent had expired. Whitney stated that on one occasion he had difficulty proving that the gin had ever been used in Georgia when there were three machines in operation so near the courthouse that the rattling of the gears could be heard inside the building.[31]

In the first six years of the 1800's saw-ginned upland cotton sold for slightly under twenty cents, while the best roller-ginned sea island cotton was bringing fifty cents a pound.[32] These prices prevailed until 1807 when the embargo disrupted trade with Europe. The differential in price was accompanied by a buyer's complaint against "false packing," a practice of placing green-seed cotton in the middle of a bag bearing the sea island label.[33] While English buyers placed a severe penalty on cotton cleaned by the early gins, which they claimed broke the fibers, the sea island cotton commanded twenty-five to thirty per cent more than green-seed cotton under the most ideal cleaning process. During this period unginned or "seed cotton" sold for two to three and a half cents a pound.[34] Thus the value added to cotton by the ginning process was more than 200 per cent. The value differential between raw and ginned cotton remained high throughout the first half of the nineteenth century.

This situation reflects the relative scarcity of gins, their costly upkeep, and their low productive capacity. It is estimated that in 1806 one farmer out of five in the upcountry owned a cotton gin. By 1820 the ratio was one to three. By the latter date more than sixty per cent of these farmers were growing cotton, but only slightly less than half of them owned gins, the average value of which was one hundred dollars each. Unginned cotton could be found stored on farms at all seasons of the year. Farm inventories in the first two decades of the nineteenth century use such terms as ginned cotton and unpicked cotton, neat cotton and field cotton, and clean cotton and seed cotton. The term "bailes" (bales) began to be substituted for "bags" about the year 1818, the new term indicating that screw-pressed cotton was taking the place of hand-packed bags.[35]

Something of the character of the early cotton industry in the Georgia upcountry is illustrated by the experience of Thomas Anderson and Bolling Anthony of Wilkes County. Forming a merchandising business near Washington in 1796, they were among the first cotton buyers in that section of Georgia, cotton factors not yet having appeared at such river ports as Augusta. They paid two and a half cents per pound for seed cotton and ginned it with their own machines. Originally they used twelve pairs of roller gins, which, with twelve operators, ginned less than than a hundred pounds a day. They later acquired a "35-saw gin" built by a local mechanic and propelled by water, which could process in one day 3,500 pounds of seed cotton. Ginned cotton at this time was selling at twenty-five cents. In June 1797, Anthony took 4,000 pounds of ginned cotton to Virginia in two wagons, paying two dollars a day for wagon, team, and driver. He sold the cotton at Bedford Court House for fifty cents a pound. For the return trip he loaded with iron at Callaway's Iron Works at Rocky Mount. The entire trip of forty-eight days netted him a thousand dollars. He repeated the venture during the following fall but found the interior markets of Virginia glutted with Carolina cotton and the price down to forty-three cents.[36]

By 1820 perhaps two-thirds of the market crops were grown within five miles of some navigable river, and much of the remainder within ten miles of some water course. Transportation was still a serious limitation to the agricultural development of the Upper Piedmont.[37] Lands lying along navigable streams sold at a premium for use in growing cotton, while those in the interior were devoted largely to subsistence crops.

The iron plow, for the most part a result of experience and inventive genius of Northern farmers and mechanics, was slowly coming into use. The types used were the cutter, the shovel plow, and the bar-share, attached to a pulling gear of twisted rawhide.[38]

In most communities the entire crop of small grain was cut with the primitive sickle. Only the more affluent farmers possessed a threshing machine powered by a horse on a treadmill which might clean five to twelve bushels in a day.[39]

Garden vegetables and fruits were not emphasized in this incipient Cotton Belt of the eastern Piedmont in 1820. Tomatoes, known then as "love apples," were appearing only on the tables of the more sophisticated farmers. They were still considered poisonous in most Georgia communities and were grown only for ornament.[40] Irish potatoes were yet unknown as a winter food on Georgia farms, although they might be bought on the markets of Augusta and Savannah where they were obtained from Northern growers.[41] The peach was perhaps the favorite fruit, but produced on seedling trees this fruit often was hard and bitter. The trees were relatively free from diseases, however, and they had long been acclimated to the environment. Since orchards often were planted to provide forage for hogs, the fruit seldom was permitted to rot on the ground and this situation kept the peach borer under control.[42] A few apples were grown, also from unimproved native seedlings. The woods and fence corners abounded in native grapes, and some foreign varieties were cultivated in the gardens of a few enterprising planters.[43]

Many of the agricultural superstitions of the late medieval period still lingered on the Georgia frontier as late as 1830. Some of these have been attributed to German and other non-English farmers who moved in relatively large numbers through the water gaps of the back country to the new lands of Middle Georgia. The Germans continued their reputation as thrifty and successful husbandmen, but their farming activities often were determined by lunar direction. All tap-rooted vegetables, for example, were planted on the waning moon, while others were planted on the waxing phase. Apples harvested just before a full moon were thought to be highly susceptible to rot. The Germans believed also that the periodic appearance of locusts from the ground was designed by Providence to enable rainwater to penetrate the earth.⁴⁴ They were careful to use compost derived in part from a stud horse on ingrafted fruit trees, and they believed that a propagation made by incision would live no longer than the parent stock.⁴⁵ An enlightened and scientific agricultural outlook awaited the development of a more mature economy and a more settled population.

Middle Georgia was firmly established as a cotton-growing area by 1830, at which time Georgia had taken the lead from South Carolina as the principal producer of the staple.⁴⁶ While only the better cotton lands were selling for as much as ten dollars an acre, Negro slaves were in great demand. Probably because slaves were poor ox drivers, the Tennessee mule was coming into use to complete the agricultural trilogy of the ante-bellum era: cotton, slaves, mules. The mule's capacity for hard work and longevity under plantation management was being recognized throughout the Cotton Belt.⁴⁷

Cotton began to show a great fluctuation in price after 1820. From 1824 to 1825 the price fell from twenty-one to twelve cents, and from the latter date to 1830 the New York price fluctuated between nine and twenty-one cents. Prices were lower for this decade than for any previous period.⁴⁸

Distress resulting from the falling price of cotton was accentuated by a series of poor crops caused by drought which began in the summer of 1827, a condition which also limited food crops.⁴⁹ The *Macon Telegraph* noted in July of that year the "utter prostration" of the farmer's hopes. Many fields of corn were entirely destroyed, livestock were suffering from lack of grass and herbage, and springs of water were drying up. Cotton, more immune to the effects of drought than grain crops, had produced an abundance of blossoms but they withered and died before fruiting.⁵⁰

The following year brought some improvement in the production of cotton, but other crops were short. In 1829 farmers again complained of poor crops as well as low prices. In that season the cotton plant grew to unusual size but it was poorly fruited, and it suffered depredations from the boll worm, which first appeared in the Georgia Cotton Belt that year.⁵¹ During the winter of 1830 relief for the impoverished condition of many small farmers who had depended solely on cotton for a livelihood was the concern of local committees in several counties. Among the activities of the relief committee in Hancock County was the gift of a Bible to all destitute families in that community.⁵²

THE RISE OF UPLAND COTTON   57

There was only one excellent crop year between 1827 and 1832 inclusive, that being 1831. In this period cotton suffered less than food crops, but the reduced price of the former climaxed the first real depression in the Southern cotton industry.[53] It is significant to note that little attention was given during this crisis to measures for improving soil fertility. To the suggestion that an agricultural society be formed for the purpose of studying this problem, the editor of the *Georgia Journal* at Milledgeville remarked that "The fault is not in our soil . . . [it] is in ourselves," and nothing was done.[54]

The sagging condition of the cotton economy was emphasized by a Burke County planter who complained that "the most judicious husbandman" could not make legal interest on invested capital. He hoped that he would never again find it necessary to plant another cotton crop, "a crop that I detest," said he, "because of its admitting of no rest, or time for the improvement of a plantation."[55] He planted several acres in mulberry trees for silkworms and he began an experiment with sugar cane "of the Green and Ribbon sorts" planted in a variety of soils. A revival of silk culture was under way in Laurens, Jasper, Putnam, Baldwin, and Lowndes counties. Many planters in Burke, Screven, and Montgomery counties reported success with upland sugar cane, and an effort to grow ginger was made in the Savannah area.[56]

In February 1828, a number of citizens convened in the courthouse at Eatonton to consider measures for relief from the distressing economic conditions. They attributed much of their current grief to such human weaknesses as extravagance and buying on credit.[57] During the early 1820's Putnam County produced annually about 8,000 bales of cotton, having a net value of $180,000, against which they paid an interest on debts amounting to $25,000. Expenses in growing this crop were estimated at $149,000. In the expenses significant items, amounting to $24,000, were hogs, horses, and mules, most of which were produced outside the state.[58] The *Hancock Advertiser* in 1828 claimed that one hog drover alone carried back to his own state from the courthouse at Eatonton ten thousand dollars in cash.[59] The committee urged restraint in foreclosing mortgages and it deplored the rapid accumulation of wealth by those who controlled credit. "It is not among the least evils of the times," it reported, "that their tendency is . . . to make the Rich, richer, the Poor, poorer."[60]

Among the remedies suggested by other groups was the curtailment of purchases from outside sources, particularly food, livestock, and manufactured goods, all of which it was maintained should be produced on the plantation. The Georgia legislature asked the newly created Committee on Agriculture and Internal Improvements to investigate the possibilities of other money crops than cotton. Among those suggested were mulberry trees for growing silkworms, olives, wine, tobacco, indigo, madder (for red dyes), and the white poppy.[61] Thomas McCall reported success in the cultivation of grapes for wine in Laurens County. He was experimenting with both foreign and native vines, and claimed a profit of $160 on one acre.

In a flow of optimism, Dr. Alexander Jones of Lexington reported a successful experiment in growing seed of the Persian poppy for the making of opium, from which he anticipated an income of three hundred dollars. He emphasized its advantages over cotton in ease of harvesting, transportation, and marketing.[62]

Such was the situation in Georgia when the "tariff of abominations" became law in 1828. The anti-tariff sentiment in Georgia appears to have originated from a set of local circumstances having only an indirect relationship to the broader economic principle which the tariff involved. A more normal agricultural situation would doubtless have softened considerably the protests against the principle of protection. Governor John Forsyth in his message to the legislature in 1828 spoke of the tariff act as recklessly sacrificing the interests of a whole section of the Union "for the benefit of a class of persons recently sprung up among us, to whom grant of special favors has been improvidently made since the close of the last war."[63] The legislature sent Congress a formal protest against this act and authorized militia officers to attire themselves in homespun uniforms. The body also proposed to extend the state's credit to George Manifee in the sum of $5,000 for establishing an iron works in Jackson County.[64] The *Georgia Journal* in a sympathetic statement denounced the tariff as "exhausting its most baneful influence . . . on the South, our native land, blighting the fairest prospects of our industrious farmers, blasting the hopes of our merchants, in short, rendering comparatively valueless, the rich productions of a fertile soil, and bounteous climate, heretofore the source of wealth."[65] The trouble was no longer "in ourselves" as that editor had earlier remarked, but in those outside interests which were sponsoring higher tariff rates.

Thus it was that the agricultural problems of the Cotton Belt became identified and somewhat confused with the national tariff question. It is significant to note that "farmers' meetings" now came to be labeled "anti-tariff meetings," and what once had appeared to be incipient agricultural societies became, by 1832, "nullification clubs." Only a slight de-emphasis of agriculture was needed to tip the balance in favor of politics. A Baldwin County anti-tariff meeting in 1828 passed resolutions against using any goods produced in the "tariff states," and urged reliance upon home manufacturing. The legislature was asked to impose special taxes on livestock and manufactured articles from western, eastern, and northern states. Similar resolutions were passed in numerous communities throughout the state.[66] The *Milledgeville Southron*, noting the disposition of Georgians to favor Tennessee livestock traders over those from Kentucky and Ohio, admonished Tennessee drovers who planned to do business in Georgia to bring with them certificates of Tennessee citizenship and to certify that their products originated in that state.[67] The *Savannah Georgian* called attention to successful experiments in the growing of cane and the manufacture of sugar in that section and hoped that coastal planters would receive protection as sugar manufacturers.[68] A product labeled as "AntiTariff Castor Oil" was made by Dr. D. A. Reese of Monticello, from home-grown bean plants. Thomas Spalding of Sapelo Island, Georgia's best-known contributor to agricultural journals of that day, began a steady barrage of essays containing much scientific information upon such subjects as sugar cane, grape vines, fruit trees, indigo, Bermuda grass, Irish potatoes, rice, poppies, and long staple cotton.[69]

There is little indication that any constructive readjustment of the state's agriculture resulted from the depression of 1827-1832. The price of cotton rose after 1830, and simultaneously the tariff rates gradually were lowered. No marked effort to curtail cotton production resulted. Domestic manufacturing received some encouragement, but the processing of textiles at this time began to shift from the home to the factory, as water-powered mills developed on the streams above the fall line. Agricultural practices still were greatly influenced by frontier conditions, and farming as a science was little known. "We have not yet learned to redeem land," said a Putnam County farmer, "or to improve it by any system of manuring."[70] The abundance of land and its relatively low price were important factors in determining the direction which agriculture had taken. Fresh lands yet could be bought for trifling sums and occupied after a short journey by wagon or on horseback. As a result, there was little incentive to change the traditional methods of farming.[71]

Before the last land lottery was completed in 1832, many Georgians were realizing that the lotteries and the abundance of cheap land had brought a serious deterioration of agricultural and social conditions in the older counties of the state. A citizen of Sparta called the land lotteries "the most unjust, unequal, and immoral" of all methods ever employed for distributing public land. The effect of the abundance and cheapness of land, said he, was that it "begat a careless, slovenly, *skimming* habit of farming." He compared farming in Middle Georgia to the hunting and slaughtering of cattle for their hides alone.[72] And where the skill of the hunter was measured by the number of cattle he killed, so the skill and efficiency of a Georgia planter were measured by the quantity of land which he was able to wear out.

V

Soil Exhaustion and Emigration

By 1830 THE NEW lands of the Southwest began to beckon planters from exhausted farmlands of the older counties of Middle Georgia. Many of these now joined the steady stream of emigrants from the Carolinas and Virginia who daily passed their doors headed for the new El Dorado of cotton planters.[1] The trek of strangers from the older states into and across the middle belt of Georgia was an old phenomenon, but it was not until large numbers of Georgians joined the cavalcade that men began to analyze the causes of out-migration. A Putnam County farmer in 1834, in noting the end of the lottery system, believed that Georgia's policy of free land had wrought great disaster in encouraging soil waste. "Already, throughout the state," he said, "[were] wasted plantations, fields turned out and grown up in broomsedge." He urged the immediate necessity for changing the system of agriculture, the alternative to which was migration to distant areas.[2]

Man seldom changes his place of abode except in the hope of improving his economic welfare. The cheapness and abundance of land in new areas offered such hope and expectation to those who now found their soils impoverished by the frontier system of cultivation. Salon Robinson described this system as consisting of shallow plowing on unterraced land, up and down hill. Without the use of contour rows all the surface soil was sent down the rivers. "Probably no soil in the world has ever produced more wealth in so short a time," he wrote in reference to the central portion of the state, "nor been more rapidly wasted of its native fertility." He believed that the cheapness of land combined with its original fertility had been the curse of this region. "When the time comes that men cannot run off to the West to get new and cheap land," he said, "then will their granite hills be appreciated at their real value, and their old broom straw fields and pine barrens be restored to usefulness and covered with a healthy, happy, wealthy population."[3]

The area between the Oconee and the Ocmulgee rivers, the first to be distributed under the lottery system, was described in 1850 as "perhaps more completely exhausted than any other part of the United States." The lands had fallen into the hands of comparatively few people, some owning from two to five thousand acres. "There is nearly one-third in woods, one-third in good condition for cropping, and the remaining one-third is, as I consider . . . so much worn out that it will not pay a man in labor for its products," wrote John Farror of Stanfordville. He suggested that exhausted land be planted in pine trees to rebuild the diminishing supply of timber.[4]

Wood in many localities, such as Parramore's hill, according to Sir Charles Lyell, was worth more than the land on which it grew.[5] A shortage of wood for fencing and for construction of buildings was noted in Putnam and Oglethorpe counties as early as 1828, when farmers began to substitute ditches and hedges for rail fences. "You may yet subsist a few years longer in plenty," warned a Putnam County citizen, "but when [your woodlands] are gone you too must go. With the utmost frugality you have not more wood now than is sufficient . . . ."[6] In this county threats were made to prosecute for trespassing persons who gathered firewood indiscriminately.[7] By the middle of the 1840's there were many farms of several hundred acres throughout all parts of Middle Georgia on which little or no hardwoods remained. Complaints were made of the high price of firewood and of the growing scarcity of fencing material. The substitution of sawed lumber for logs in building Negro cabins was recommended as a conservation measure.[8] In commenting on these conditions a Milledgeville editor in 1844 cited the disappearance of original forest growth and the scarcity of fencing material as the reason for "the desertion of cherished homes and associations and a pilgrimage to new and untried fields of companionship." A few wire fences were coming into use in the 1850's and these were said to cost less than wooden rails.[9]

The short supply of timber apparently was to some extent a result of the clearing of woodland for the expansion of cotton growing in the immediate vicinity of the older settlements. David P. Hillhouse of Wilkes County stated that it would cost him three times as much to restore old land as it would to buy virgin land. His practice was to clear woodland in the vicinity of worn-out fields in the hope of attracting a buyer to whom he could sell his plantation. While he adopted a rudimentary system of crop rotation, he eschewed the tedious practice of contour plowing, because the practice was inconsistent with the plentiful supply of land and the relative scarcity of labor.[10]

The disastrous results of the vicious combination of cheap land and expensive labor were evident in all parts of the Georgia plantation belt by 1845. Southwest Georgia and the Florida Panhandle were no longer considered as agricultural frontiers in this period, nor were virgin lands in Alabama as plentiful as formerly they had been.[11] A native Georgian writing from Florida in 1837 noted that he and his neighbors had approached the limits of their peregrinations "unless we fly off in a western tangent," he said, "and it will be a long journey in that direction which will bring us to good uncultivated lands."[12]

Throughout the last two decades before the Civil War, any further out-migration of farmers from Middle Georgia was discouraged by civic-minded citizens in all walks of life. They saw in emigration a serious challenge both to stable land values and to social stability in their communities. A Hancock planter, who with a show of civic pride signed himself "A Middle Georgian," gave a bleak description in 1844 of conditions in the new lands. "Within the last fifteen years, between forty and fifty families have moved from this county and settled in Southwestern Georgia or New Alabama," he said. "More than half of these have died. None . . . have improved their estates."[13] Eli H. Baxter in a public address cited the advantages of Middle Georgia, "worn out and exhausted as it is," over the rich lands of southwestern Georgia which he described as "but painted sepulchres," full of disease and death. "Their children that survive infancy," he affirmed, "are delicate, sickly, and rickety from the cradle to the grave."[14] The Macon *Journal and Messenger* suggested that farmers on poor lands should, instead of moving, sell their surplus slaves to good masters and put the proceeds in railroad stocks.[15]

That the tide of migration to southwestern Georgia appears to have ebbed by 1850 is due perhaps to economic changes rather than to such public utterances as these. The people who had moved into Georgia's western counties appear to have been youthful and vigorous.[16] They carried with them the traditional plantation practices of the older areas, which practices continued without the modifications to which they were subjected in the decaying communities of Middle Georgia. By contrast, the newcomers were influenced by a frontier spirit of expansion, and by visible signs of increasing prosperity around them. As a result, by 1850 there were in the newer areas many evidences of the same kind of soil exploitation which had impoverished the older counties. Alexander McDonald, in 1845, described most of the farmsteads in the vicinity of Columbus as presenting anything but a thriving appearance. "Many of our farms appeared to the passerby as if a shower of rails had fallen on one day and a shower of houses the next," he wrote.[17] Two years later Sir Charles Lyell observed that the Chattahoochee River at Columbus had begun to resemble the Oconee at Milledgeville where the clearing away of the woods and the plowing of the watershed had converted the stream into a torrent of mud and silt. "I am assured that a large proportion of fish, formerly so abundant in the Chattahoochie [sic] have now been stifled by the mud," he wrote.[18] A native of that section described much of the area drained by the lower Chattahoochee as beginning to "put on the wrinkles and furrows of premature old age brought on by a soil-killing culture."[19]

In 1851, John Trott of Troup County summarized the agricultural situation in western Georgia, less than twenty-five years after it was opened for settlement. He wrote:

> We are awfully bad off up here, having nearly worn out one of the prettiest and most pleasant counties in the world. You would think it a gross slander, when politicians sometimes say "Old Troup," I have no idea they intend any disrespect. . . . It is true we have many of the marks of age, and if a young Rip Van Winkle should find himself suddenly waked up in the middle of some of our large plantations, when he looked out upon the waving broomsedge, the baren hillsides, and the terrible big gullies, it would not be wonderful if he should feel that he was about home. . . . We have felt that things were not as they once were, and that the prospect ahead was full of gloom . . . "What shall we do?" The most general conclusion and the one most in conformity with Georgia usage was, to move to a new country. This, however, was easier said than done . . . [We are] a long ways off from the wilderness. And then here are our good homes, good society,

good health, good schools, and soon to have a good railroad . . . We can do very well if we husband all our resources. But when we pay for horses, pay for mules, pay for pork, then for a few axe halves, and brooms, and plow stocks and sundry other Yankee notions . . . out of a worm-eaten cotton crop . . . The pile which I can call my own . . . [will decrease].[20]

At the same period a resident of Pine Mountain Valley in Harris County wrote with nostalgia of an earlier day when farms were rich and productive. "But alas!" said he, "these days of prosperity are gone. The fields . . . are reduced to sedge grass, all scarified with gullies."[21]

Farther north, in Carroll and Heard counties on the border of Cherokee Georgia, the situation appeared to be more encouraging, probably because of the high preponderance of small landowners in that section and to the relative scarcity of Negro slaves. A more favorable situation also prevailed generally throughout Cherokee Georgia, except in its more fertile valleys where there were lamentations over a soil which had lost its original bloom. Yet the sound of the frontiersman's axe could still be heard in the less desirable valleys and coves and on the ridges of that last agricultural frontier in Georgia. A Floyd County farmer claimed in 1861 that the natural fertility of the valley lands of the Coosa, Etowah, and Oostenaula rivers, instead of inspiring agricultural reform, had by their unusually high productivity "stultified the spirit of enterprise."[22] Charles W. Howard, a resident of the region and one of the foremost agricultural writers of Georgia, cautioned farmers in the new and fertile sections to guard against the follies of planters in the older lands. "Twenty years have elapsed and some of these fine lands are nearly exhausted," he said. "We know land in Cass County, which twenty years since produced fifty bushels of corn to the acre, which will not now produce twenty bushels." He also pointed to exhausted farms in Floyd and Gordon counties where some farmers were able to make a living only by selling slate found on their land.[23] The *Southern Recorder* noted editorially that "even in Cherokee Georgia" with its rich valley lands and "beautiful hills and mountains, the Vandal spirit . . . is already rioting. . . . Scarcely a quarter of a century has passed away since civilization found a lodgement in that beautiful and fertile region and yet [the residents of] the rich valleys of Cass, Floyd, and Chatooga, are already contemplating the gradual decay of their enclosures with something like alarm."[24] A Walker County farmer analyzing the status of agriculture in the northwestern portion of Cherokee Georgia wrote in 1853:

> Our farmers trust too much to their rich, fresh lands ... Corn one year and wheat the next, with a little oats occasionally—is the most universal practice of our farmers. Manure is never thought of .... Their stock roams at large in the woods .... In short, they are farmers of the old school. The idea never entering their heads that their's is an exhausting ... system. Already the broomsedge begins to wave over some of these beautiful lands ....[25]

The census returns for 1850 and for 1860 show that Cherokee Georgia experienced only a slight increase in population during that ten-year period, and the increase was largely in the more mountainous counties. The more fertile areas failed to gain perceptibly, and some, notably Cass and Chatooga counties, actually lost population.[26] Ralph King, a Georgian at the American consulate in Bremen, Germany, attempted to direct German immigrants to Cherokee Georgia in the 1850's but he had little success.[27] Daniel Lee, editor of the *Southern Cultivator,* thought Northern farmers might be induced to settle in Cherokee Georgia to take the place of those who were emigrating. The region, he said, presented "the most inviting field to rural industry and enterprise to be found on the continent." However, he cautioned that

> The Northern man should have nothing to do with legitimate planting. Leave that to the worthy citizens who understand the business; but know nothing about wool-growing, making butter and cheese, nor any mixed agriculture. We are paying two dollars a bushel for Northern Irish potatoes. ... We have fine apples from Knoxville.
>
> Northern hay sells here at twenty dollars a ton, when it can be raised at one fourth the money. One can buy fair farming land ... on the railroad toward Nashville at from two to five dollars an acre. These lands have some buildings and fences upon them. They belong to planters who want to emigrate with their "force" to the richer virgin soils of ... Arkansas and Texas. This tide of planting emigration, setting so constantly westward, creates a vacancy for a new race of legitimate farmers. The latter do uniformly well; for they get their land for a song; they have an excellent market at their doors, and purchase ... as cheap as farmers of Western New York do.[28]

The emigration from Georgia appeared to reach its peak during the middle forties. At Milledgeville, in 1846, Sir Charles Lyell heard the Presbyterian minister exhort his congregation "to take the same view of their sojourn on this globe, which the immigrant takes on his journey to the west," knowing that his troubles would soon end in a happier land.[29] General financial stringency ushered in that decade, and in 1842 the price of cotton reached the lowest point that men had ever known. From 1840 to 1848, when conditions began to improve, the price ranged between six and ten cents.[30]

The general distress was deepened by poor food crops during much of that period. In the early summer of 1843 the wheat crop was badly damaged by rains, and during the following winter severe cold caused the death of livestock in some parts of the state.[31] In 1845, a year in which the heaviest emigration was reported, a drought in most of Georgia caused the failure of provision crops.[32] Distress and discouragement were the common lot of both the rich and the poor. Garnett Andrews of Wilkes County wrote in this period that several hundred acres in Middle Georgia often were sold for less than a dollar an acre. "The usual rule is to sell the woodland for what it may be thought to be worth, and give the purchaser the old lands and the houses for nothing," he wrote. He stated that a thousand to fifteen hundred dollars would buy a comfortable house with garden and buildings and several hundred acres of land.[33]

Farmers who had made substantial improvements on their plantations seem to have been less likely to emigrate than others.[34] "I have built me a comfortable house, my negro houses, barn and stables, are all good, I have five orchards, my wife has beautified our yard with flowers and shrubs [and] we have an excellent set of neighbors," wrote a Middle Georgia farmer who refused to exchange these advantages "for a comfortless cabin in the West."[35] A study of land values of individual farmers in Hancock County in the 1850's indicates that the more productive lands were in the hands of wealthy owners who apparently had no desire to part with them. These owners were the larger planters who had secured a great proportion of unimproved land available in the vicinity. This was like buying insurance against the necessity for moving. The unimproved land was used for pasturage, the timber on it was cut for fuel, and it was brought into cultivation when the older fields became worn out and unproductive.[36]

In the minds of many, poor houses were associated with soil waste and emigration. "They seek fresh lands and another cabin," wrote Charles A. Peabody, who believed that hallowed associations surrounding a comfortable house were excellent antidotes for emigration.[37] The size and appearance of a planter's house were no measure of his wealth, however, as Emily Burke discovered during her sojourn in southeastern Georgia. She found the possessions of planters to consist mainly of slaves, livestock, carriages, furniture, and plate which, said she, "could be transported when the occasion demands a removal from one old worn out plantation to another."[38]

Indeed, the planter's house of the 1830's was generally described by travelers as simple and not always substantial. "[All] houses that you meet with on the road-sides of this country is two square pens, with an open space between them, connected by a roof above and a floor below," wrote the English traveler James Stuart in 1833, in describing the area between Macon and Columbus. He noted that the kitchen and slave cabins all were separate buildings, as were the stables and barns. About ten buildings of a simple description made up the establishment of a planter who owned half a dozen slaves.[39] The proximity of the Cotton Belt to the frontier was a significant phenomenon in the history of Georgia agriculture throughout the ante-bellum period. Even as late as 1851 ex-Governor John Forsyth described the planter's house generally as a rude structure made of logs "with rail fences and rickety gates" guarding its enclosures.[40]

The desire to emigrate was by no means limited to men of modest means, however, for adventure, enterprise, and economic opportunity appealed alike to yeoman and planter. Also the factor of risk involved laid a restraining hand upon the timid and unambitious as well as upon those who had a heavy stake in the older communities. The larger planters who emigrated were also required to make costly and painstaking preparations. William Lumpkin, who moved from Clarke County to Mississippi in November 1837, sent his son ahead in the previous spring to plant a crop and to prepare for the arrival of the family. Previously this son had gone to Mississippi to select land and to make the purchase. Before departing, the elder Lumpkin had his notes discounted and used part of the cash to settle his debts. He disposed of his furniture and similar belongings, hoping, as he said, that it would never be necessary for him to return to Georgia. "We shall have upward of 60 Negroes, 32 horses & mules, almost 50 head of Cattle besides the white family, consisting of your mother & myself, 4 daughters, 4 grandchildren, Mr. Mayer, & I, and perhaps 2 young white men," he wrote to his son. "I have a Coach & 4 white horses, Two large waggons, & my sulkey—Mr. Mayer has 2 large waggons a Coachee and cart & 14 head of horses and mules—so you see we shall have a very large cavalcade, & a very expensive one...."[41]

Such a caravan might travel twelve to fifteen miles a day, but Lumpkin was optimistic in hoping to arrive at his destination by Christmas. He found the roads were makeshifts, bridges were almost non-existent, and ferry tolls were high.[42]

An English traveler in this period described the route of migration from Columbus west into Alabama and Mississippi as filled with an "almost uninterrupted line of emigrants with ennumerable [sic] heavy and light wagons." Some of the vehicles were stuck fast in the mud and the men around them were pulling, whipping, and swearing. "At one time of the day we passed 1200 people, black and white, on foot," he wrote. There was also a more affluent class of emigrants who traveled in light carriages, among which he observed an old-fashioned gig pulled by a horse guided by an aged grandmother, with several white and black children stuck in around her.[43]

The most popular route used by these emigrants through Georgia was from Augusta through Sparta, Milledgeville, Macon, Talbotton, and Columbus. From Columbus the old Federal Road continued through the Alabama Creek country west to Natchez. The path of migration ran the entire length through the heart of Georgia's Cotton Belt, and it was largely from this region and its fringes that the local tide of emigration flowed. The emigrants, having originated in the Piedmont areas of Georgia, the Carolinas, and Virginia, sought a region whose topography and soil were familiar to them. Emigrants from the foothill and mountainous areas of the upper South moved along more northern routes which brought them into and across the southern portion of Cherokee Georgia. Richard Peters came to Georgia from Pennsylvania and selected a site for a farm in Gordon County because it reminded him of his native Chester County where topographical and geological formations were similar to those in his newfound home.[44] Sir Charles Lyell, with the trained eye of a geologist, was quick to note the relationship of geography to patterns of migration. "[They] who go southward from Virginia to North Carolina or South Carolina, and thence to Georgia and Alabama, follow, as if by instinct the corresponding zones of country," he said. "The inhabitants of the red soil of the granite region keep to their oak and hickory, the "crackers" of the tertiary pine barrens to their lightwood, and they of the newest geological formations in the sea islands to their fish and oysters."[45]

Lyell noted that the emigrants on reaching Texas found their bearings were grievously at fault. The unorthodoxy of the flora and soil in the state often resulted in faulty judgment concerning the quality of the lands there. Pine trees did not grow abundantly west of the Sabine even on poor land, and tree growth no longer served as an infallible criterion for the quality of soil.[46] The fertile black lands of Texas were uninviting to many, and when encountered in wet seasons they were positively forbidding. Andrew J. Prior, writing from Rusk County, Texas, to a former neighbor in Georgia, enthusiastically described land which he had just acquired, in terms of the topography around his old home near Cedartown. "The Color in general [is] about the color of them yellow whetstones that Mr. Bigelow has in his store . . . representing sand and as fine as our Cedar Town land I cannot call it anything else but sand," he wrote, "and for further explanation it is exactly like the little hill after you come out of the creek cumming to Cedar town at the Buck hannon ford . . . ." He described the trees as similar to those in Cedar Valley; the woods were full of grapevines, but the water was bad. "The face of the country in general lies in Waves not to say brokener than from your house to where I lived at the Lacy Witcher place . . . ."[47]

Apparently Prior did not prosper in Rusk County and spent only a year there. He then bought land in Sabine County, and again not only ran the hazards of poor judgment but fell into the hands of unscrupulous land agents. His son, showing more candor perhaps than his father would have admitted, wrote to his friend in Cedartown as follows:

> In this country [it] is mean as smelling the pole cat in the woods. The people tell many lies about the land & water . . . . They are fools & rascals & great liars of flattering for the sale of the land. Many of the people left Texas as when [we] saw by the road . . . . I will leave this mean country an its mean smellings an I will run to Georgia by myself . . . [My father] had no judgement & follishness of purchasing the poor land. The cotton is poor & little heads . . . . I will not live here in the summer because it will be sick mostly the water will become green an I hate of drinking it . . . .
>
> You and your friends of Cedartown must not sell your land . . . which is very good . . . as I hate to deceive them of their mis-

takenings of movement to Texas from the best county in Georgia.
I am pretty sick usually. My belly was often pained with ache & pucked often. I am very lean & pale....⁴⁸

The migration of Georgians to Texas and the far Southwest was heaviest in the last decade before the Civil War, the earlier migrants settling largely in Alabama and Mississippi. In 1850 there were 59,000 natives of Georgia living in Alabama and 17,000 in Mississippi while fewer still were in Louisiana, Arkansas, and Texas. The greater the distance between Georgia and the receiving state, the less was the migration from the former. However, more than twice as many Georgians moved to Texas in the decade of the 1850's than had gone there in the preceding thirty years together.⁴⁹ This migration to Texas was the final stage of Georgia's out-migration, and it drew most heavily upon the newer counties in the western and southwestern parts of the state. Charles A. Peabody, conducting a tailoring establishment at Columbus, complained that he was forced out of this business and into farming partly because his customers in large numbers had left for Texas with unpaid bills.⁵⁰

Soil exhaustion in the western counties was an important factor, although not the only one, in the emigration from Georgia in the 1850's. It was a decade when cotton planters generally were making satisfactory profits, and slaves as well as cotton were commanding relatively high prices.⁵¹ Some degree of prosperity had returned to the older region of Middle Georgia after the depression of the early forties. The more successful planters had increased the size of their holdings and improved their managerial procedures. By 1860, Georgia as a whole had resumed its normal growth, having received over seventy-five thousand more people in the previous ten years than had emigrated.⁵²

Just as earlier those in the older counties had by propaganda attempted to counter the out-migration of their people to the new Georgia counties, in the 1850's Georgians in the western part of the state were ridiculing what they now called "the western fever" which was attracting their own people to the trans-Chattahoochee region. Perhaps the best known of the self-appointed propagandists was the poet Francis Orray Ticknor, who lived near Columbus. One of his poems, "The Mover Man," was based upon the adage that a rolling stone gathers little moss.⁵³ Others were no less diligent in their crusade against emigration. A Troup County planter offered an antidote for the desire to emigrate. It consisted largely of a method of horizontal culture or plowing on contours, even though the practice involved devious long rows and many short ones. "I have had the western fever," he admitted, "and visited more than half the counties and Parishes in the States of Louisiana, Mississippi, and Texas, but nowhere have found a place superior . . . to the home of my boyhood. Many of my relatives have emigrated to the 'West,' they have not done better . . . and have undergone many privations." He pointed out that the good health, educational advantages, and "good society" enjoyed in the older states more than compensated for the fertility of Western soil.[54]

"Why . . . should we, like a set of dastards, desert the graves of our fathers, in pursuit of a virgin soil?" wrote another farmer. "I, for one, am determined to hide my red hillsides and fill up my gullies. If I succeed in doing this, I shall have an additional cause for consolation on my death bed."[55] Much publicity was given to a news story then current concerning a man who ran off to Texas to evade trial for murder. He returned to Georgia six months later, declaring that he would rather be hanged in Georgia than to live in Texas.[56]

Emigration from the newer western and southwestern Georgia counties never quite reached the startling proportions of the earlier movement of farmers from the older belt of Middle Georgia. Only nine of the fourteen counties making up the total area of the last two Creek cessions showed a decrease in population over the ten-year period ending in 1860. The five which gained population in this decade were Carroll, Heard, Coweta, Campbell, and Lee. With one exception these five counties are in the northern part of the area which was ceded in 1826. Two of them, Carroll and Heard, lie in a triangle between the Chattahoochee River and the Alabama line north of West Point, a region of foothills and yeoman farmers. The nine counties of Harris, Marion, Meriwether, Troup, Talbot, Muscogee, Randolph, Stewart, and Sumter, in the central and southern portions of the area, had 17,346 fewer people in 1860 than they had ten years previously.[57]

VI

Diversification and Economic Independence

AGRICULTURE in the Georgia Piedmont began with the huntsman and the cattle drover, who in turn were replaced by subsistence farmers.[1] By 1820 cotton as a staple crop had come into prominence in this area, and it came to mould the social and economic pattern of the entire region. As a result of the depression of the early 1840's the cotton growers of the Piedmont returned briefly to subsistence farming, but with the refined modifications of a society now somewhat removed from the frontier. During the decade of the 1850's, economic nationalism with its mixed-farming enterprises waged a losing battle against the resurgence of King Cotton. If the decade beginning in 1840 may be thought of as one in which economic independence was a paramount consideration, the following decade was one characterized by the development of Southern nationalism. The bid for political independence in the 1860's was therefore preceded by a strong movement to achieve economic independence. In agriculture the latter movement was manifested largely in an attempt to diversify crops and to introduce new ones. Along with diversification of crops and the introduction of new ones there was a mild industrial revolution.

The trend toward crop diversification first became apparent in the wake of low cotton prices and the tariff controversy between 1826 and 1832.[2] The retreat from cotton in this period, however, was generally in the direction of breeding livestock, with the emphasis upon cattle and hogs, rather than in the direction of a more general program of diversification. "Let those farmers who are free from debt, devote their attention forthwith to the raising of Cattle of every description, and abandon the cultivation of cotton," said the *Hancock Advertiser* in 1826. "Let those who are in debt, divide their attention between the raising of cotton and cattle, and make their retreat from the former, and their advance into the latter gradual but determined."[3] The paucity of agricultural leadership at this time and the relative absence of agricultural literature and organization precluded the development of a more well-rounded system of diversification. Knowledge of the culture and processing of new crops was not easily acquired or disseminated, and anything resembling a genuine agricultural revolution was looked upon with serious misgivings.

David P. Hillhouse, whose planting operations were in Wilkes County, appeared to be a recognized spokesman on agricultural readjustment in the earlier period. He thought it unwise for a farmer to attempt raising everything that he needed. His own practice was to exchange beef and wool for the homespun cloth and flour of a neighbor. He pointed out that "Many persons have talents for practicing with success in a few branches of agriculture, which they smother . . . by efforts to comprehend all its parts." His own attention was confined to corn, cotton, and some livestock.[4] A haphazard division of labor within the community was characteristic of his region. Something resembling a division of labor between whole communities developed in later decades with the growth of a more highly organized society.

Hillhouse's theory of a division of labor within the local community was impracticable because of the small farmer's tendency to imitate the practice of his more affluent neighbors. Both groups planted their best land to cotton, with corn as the most important secondary crop. All remained essentially cotton planters and none was able to buy from his neighbor the subsistence products which he required. In explaining his own devotion to a single large crop of cotton in 1829, Hillhouse said: "Now, if I had undertaken to make a little sugar, and a little wine, some wheat, potatoes, oats, flax, indigo, tobacco, feathers, and pindars [peanuts], my hands would have been throughout the year dribbling and jobbing, and there would have been little to show for their labor by May-day."[5] It may safely be assumed that his neighbors, who worked their lands with little or no assistance from slaves, came to the same conclusion.

One new agricultural enterprise appeared in the Piedmont as

a result of the 1826-1832 depression. This was a brief and somewhat desultory revival of silk culture which, combined with home manufacturing, seemed to offer some promise of relief. The revived interest in silk was not confined to Georgia or even to the South. Many experiments with silk occurred in Virginia before 1830, where farmers hoped to devote their worn-out tobacco lands to its culture and thereby check emigration. Through the *Farmers' Register* interest trickled southward into Georgia where the enterprise was described as one which not only would supplement income from other pursuits but give employment to the young, the aged, and to infirm Negroes.[6]

The lingering vitality of the silk industry received new stimulation by the depression of the early forties. Its most persistent defender in Middle Georgia was Asa E. Earnest of Bibb County, who claimed consistent and successful results for many years. "The condition of our country certainly requires that we should all be doing what we can to relieve it from its present unhappy state," he wrote during the depression of 1842. "If I can throw $1,000 or $10,000 worth of silk into market every year, I shall do something [for my country]."[7] Richard S. Hardwick, who entered the business around 1840, believed the industry might be profitable if machines could be procured for preparing the silk for market.[8] In 1847 Thomas Holley Chivers, who raised a million silkworm eggs that season, reported the invention of a silk-winding machine, "spinning it into thread, doubling and twisting it all at the same time."[9] His invention seems to have been stillborn, however, for Georgians continued to experience difficulties in processing and marketing the product. There was no local market for cocoons and raw silk.[10] The industry appears never to have advanced much beyond the stage of a novel home enterprise.[11] One of its principal results was to revive an interest in the history of the silk industry of the Colonial period. By 1845 the revived industry had practically disappeared despite legislative subsidies to stimulate it. In 1859 less than fifty Georgians applied for the legislative bounty,[12] and the census for the following year recorded relatively small quantities of cocoons in only thirty-five counties.[13]

The new silk industry was already waning when the financial stringency of the early 1840's brought on the second phase of agricultural diversification. It was in this period that the agricultural reformers played an important role in directing the course of agricultural methods and practices to achieve economic independence. The logical course of economic self-sufficiency was toward Southern nationalism and political independence which came at the end of the ante-bellum period.

Interest in humbler crops than cotton was stimulated by extremely low cotton prices at the beginning of the 1840's. Farmers began to see the wisdom of producing at home those commodities formerly purchased from merchants.[14] Over-attention to cotton was blamed for deficiencies in the quantity and quality of fruit, vegetables, milk, and other items which might add comfort and health to plantation life.

Those who advocated diversification in this period emphasized the full life in a rural environment as the ideal and goal of the new movement. "The pursuits of agriculture have become not a mere business of dollars and cents . . . but a business of pleasure," said Richard P. Sasnett in an address before the Hancock Planters' Club in 1846.[15] Others wrote eloquently on the luxury, beauty, and contentment to be found on a diversified plantation. "The neigh of the horse, the low of the cows, the bleat of the sheep . . . the scream of the pea fowl [raise] their chorus, and there's life for you!" said one. "But how lank everything looks when all is cotton, corn cribbed in a small pen, [and] wheat in a goods box!"[16]

The larger planters with heavy investments in labor, many of whom had scattered plantations under the care of overseers, did not find diversification practicable. Such a program was incompatible with the specialized skill of the average farm overseer and with the ability of the large working force required. Absentee planters therefore bore the brunt of much ridicule. One critic compared their tastes and habits to those of Northerners, and he condemned both groups for their "foolish and disgusting fashion of loitering for weeks and months" at famous watering places. "We would as soon be seen with our family at one of those fashionable brothels, in our great cities, as at some of these haunts of vice and criminality," he said. He condemned the planter who eschewed rural life and "aspired to no higher abode than a city mansion," attempting to live in the improper fashion of urban society.[17]

The diversification movement was not only critical of absentee farmers and their penchant for urban fashions, but extended the criticism to include also the cultural and social habits which the one-crop system imposed upon rural dwellers. Critics noted the absence of shade trees around Negro cabins, with cotton planted up to the door. The farmer himself was described as living in an uncomfortable house set in the middle of a cotton field. "I stopped once at the house of a friend who worked some 25 or 30 hands, a gentleman of education and refinement, with a nice and accomplished wife," wrote an abserver in 1843, "and they put me to sleep in a room . . . half filled with cotton, and without a fireplace."[18] So completely was this planter's regime geared to cotton that he used no servants in the house but employed them all in the fields.

Among those who proposed the new order of diversification was Professor Charles McCay of the University of Georgia, who cited economic theories supported by figures to show the advantages in producing wool instead of cotton. "The former would be worth forty cents on the plantation at the lowest," McCay said, "the latter not to exceed twelve cents, which makes a difference in favor of wool-growing of more than three hundred per cent."[19] These critics took delight in relegating the role of "King Cotton" to that of serf. It was pointed out that, because of its low price, cotton was replacing Spanish moss, animal hair, straw, and corn shucks for mattresses used in slave cabins.[20] Beds made of "live geese feathers" from Tennessee were certainly not superior to those made of cotton, reasoned one.[21] New uses for cotton proved encouraging to growers, however. A roofing material with a cotton base was among the successful experiments reported.[22]

The de-emphasis of cotton was a concomitant of the economic nationalism prevalent throughout the South during the period following 1840. Because cotton was chiefly an exportable commodity, it entailed upon the grower both the necessity and the obligation to purchase food, plantation accoutrements, and manufactured articles from outside sources. De-emphasis on cotton therefore fitted neatly into the evolving movement toward Southern nationalism. Typical of the Southern nationalistic mood was that of a Georgian who, in advocating a live-at-home program, concluded his discourse by saying: "I am sick and tired of being 'hewers of wood and drawers of water' any longer for the North."[23] A citizen of Habersham County expressed the mood more clearly when he wrote in 1850 as follows:

> It cannot be doubted that our state possesses all the natural resources for true independence. Let us, then, show the patriots at Washington, who are discussing the difference between working a black mule and a white ass that their decision is of but little importance to us. We will teach those Yankees who are ever meddling with our institutions, that some things can be done.

> ... Let us resolve to be an independent people. This must be accomplished by pursuing the Home System. Let us raise everything we eat whether it be of animal or vegetable origin. ...
>
> On the banks of the Chattahoochee and Ocmulgee, we will build our Lowell and Manchesters. ... Thousands of our people might be profitably employed in manufacturing the great staple of Georgia into yarns and cloth.[24]

He cited Georgia's possibilities in producing rice, wheat, corn, rye, barley, sugar cane, oranges, and lemons, as well as hay, and livestock. "Let our men of the soil awake from their Rip Van Winkle sleep—be true to themselves—and we can and will be an independent people," he concluded.[25]

It became fashionable to discredit the virtues of all imported goods and to emphasize the merits of home productions. Commodities brought into the state from the North were likely to be proclaimed as spurious. Northern fruit trees were declared unsuitable for the Georgia climate, and Ohio bacon less nourishing than that grown in Georgia. Both livestock and seed bought from Northern dealers as "certified stock" turned out to be of inferior quality, it was claimed.[26] Oriental figs were too filthy for the tables of civilized Southerners. "The world does not produce finer figs than these [in the] Southern States," said a writer in the *Southern Cultivator*, "and they can be prepared for keeping and for market just as easily as in Asia."[27] The excellence of okra soup was not neglected and the famous gumbo soup of New Orleans became widely celebrated.[28]

Farmers who had sustained losses in acclimatizing new livestock breeds, or who had been victimized by unscrupulous salesmen, found it easy to join the ranks of the economic nationalists. One writer estimated that the people of Putnam County in 1844 paid five thousand dollars for pork alone, not to mention "carriages, horses, and mules ... tubs, pails, buckets, axe-handles and axes, hoes, fine boots and shoes, negro shoes, saddles, bridles, etc."[29] Richard S. Hardwick made a countywide census of Hancock County in 1845 to determine the cost and quantity of commodities bought from the outside "in order to demonstrate the causes of the country's distress." The practice of giving away silver cups as premiums by the agricultural societies was deplored. "They are not made by our own mechanics; and the money used to purchase them, therefore, has to go North, thus adding to the drain upon our resources and industry that has brought the South to what it is."[30]

It was claimed that twenty thousand tons of paper annually were shipped to the South from New England mills, representing a tremendous drain on the section's cash resources. "[Yet] South Carolina and Georgia could produce material for paper to supply the world," said one. He also suggested the use of rice straw for making newsprint.[31] To set an example of patronizing this new industry, the *Southern Cultivator,* in 1849, printed its November issue on paper made at the mill of Chase and Linton in Athens. Various newspapers in Georgia followed this example.[32]

A host of new crops and a few novel enterprises were advocated with the vigor of a political campaign. For planting on poor land the Jerusalem artichoke was recommended as a crop to provide food for hogs and sheep.[33] The Osier willow could be grown for making baskets, while Salsola might be grown on the salt marshes of the coast and used for making soda.[34] It was claimed that one acre of cultivated beargrass would produce five or six tons of "Southern Hemp" processed in the usual simple manner of other types.[35] A merchant reported seeing some cotton baled with Georgia beargrass cordage which "could not be distinguished from the regular manilla." Okra fibers also were considered as a substitute for hemp, while the seed of this plant were recommended as a substitute for coffee as a beverage. "Its ripe seeds burned and used as coffee cannot be distinguished therefrom," wrote a planter, "and many persons of the most fastidious taste have not been able to distinguish it from the best Java."[36]

In the quest for new crops and new uses for old ones, many farmers fell prey to dishonest vendors of spurious commodities. "There is no pleasure which . . . [farmers] so reluctantly forego," wrote one, "as that of being humbugged."[37] One of the more famous agricultural humbugs apparently had its origin in Georgia in 1835 when Amos W. Hammond of Ruckersville wrote to various newspapers advocating the planting of "Florida coffee." He claimed it would produce high yields under simple methods of culture, and that it would not be eaten by insects and predatory livestock.[38] Wyatt Allen of Twiggs County was among those who purchased and planted some of the seed. After harvesting the crop, according to his own admission, he prepared a beverage from the product and barely escaped death from its poison.[39] This plant apparently became distributed over a wide area, for two years later a North Carolinian was informing the public of the true character of what he called "that offensive weed." He admitted that the two dollars he paid for a pound of seed and the labor of cultivating an acre of land were not half so mortifying to his pride "as being made the dupe of a Knave."[40]

Charles W. Howard was among the few who called attention to the South's economic vassalage to the West. "We speak of dependence upon the North—we are more hurtfully dependent upon the West," he urged. While the South bought "Yankee notions," the Yankees at least bought cotton in return, while the hog drover from the West carried home only specie. "We defy anyone to produce an instance of a country whose agriculture was permanently prosperous, which bought its meat and bread," he said. He declared that pork was the most expensive meat that Southerners could consume, and that the eating of "Yankee butter" was an unpardonable sin.[41]

Pork in the diet of Southerners also was condemned by Dr. John Stainbach Wilson of Columbus who conducted a health column in *Godey's Lady's Book*. "It is full of impurities, it is often partly spoiled, and it is . . . heavy and indigestible," he wrote. Frying was the "most abominable" of all methods of cooking. "If Southerners cannot manage to take in a sufficient quantity of grease, in the form of fat bacon or pork . . . they are sure to supply any deficiency by saturating their peas, beans, potatoes, bread, and everything else, in . . . hog's lard," he complained.[42] Combining professional advice with patriotic fervor for economic independence, another physician advocated wooden shoes for Negroes, claiming that wearing them would "unite the acceptable qualities of cheapness, warmth and dryness," and would prevent consumption.[43] Robert Nelson cited the general use of such shoes among the farmers of northern Europe. He named a number of native trees such as willow, cottonwood, basswood, poplar, birch, and others of indigenous growth which were ideally suited for making shoes.[44]

Nelson, a horticulturist by training and a native of Denmark, also professed great admiration for Southern flowers and he did not neglect to cite the place of indigenous plants in this campaign for self-sufficiency. He was ridiculed for placing in his nursery at Augusta such common flowering plants as prince's feather and cockscomb. "We have thrown away a heap of fine old plants (merely because they were old fashioned), which we ought to get back again, while we have introduced many new ones whose only merit is to be new," he retorted.[45] Charles A. Peabody believed that many excellent Southern fruits were neglected simply because they were commonplace. He thought that the maypop, or passion fruit, which grew on every red ditch bank, was superior to the pomegranate and "equal to the best Cuba orange.... Let us improve the fruit adapted to our soil and climate first, and then acclimate others," he insisted.[46] The persimmon was called the true American date. "It is a very palatable fruit, and might become, under high cultivation, a very delightful addition to the dessert," said Stephen Elliott, Bishop of the Georgia Diocese of the Episcopal Church, in an agricultural address at Macon. "I venture to affirm, that if our eastern friends possessed it . . . we should soon hear of it as large as a peach, and receive it in rich abundance as a substitute for dates and dried figs."[47]

While new crops were being advocated, new uses for old crops were discovered. A planter in southwestern Georgia discovered that one acre of old land would produce over a hundred bushels of peanuts and that these were superior to corn for fattening hogs.[48] The chufa, called the "earth almond," was introduced as excellent food for both man and beast, but its resemblance to the tenacious nutgrass invited suspicion.[49] Upland rice became a common crop on the bottom lands of western Georgia, where productions of fifty bushels to an acre were reported.[50] A Texan called attention to the possibilities of the pecan, but it was not until a later period that this nut made its appearance in Georgia.[51]

While the Georgia legislature had appropriated little money to enable geologists to discover the state's hidden resources, men with scientific training and a zeal for promoting extractive industries were calling attention to the untapped mineral wealth of the state. Dr. Edmund Monroe Pendleton, a planter-physician living in Hancock County, mentioned in 1848 the discovery of diamonds in Hall and Gilmer counties, and cited the existence of gold deposits and marble beds in other parts of North Georgia, and also valuable deposits of copper, iron, pyrite, saltpetre, kaolin, blue and shell limestone, and coal. "No one knows how valuable these deposits may yet prove in reclaiming the worn-out lands of the cotton region of Georgia," he stated.[52]

Farmers in many sections of the state already were finding profitable employment for labor in various extractive industries. Before 1860 Zadock Bonner, by using slave labor and crude methods during slack seasons, mined more than a quarter of a million dollars' worth of gold on his plantation in Carroll County. Other family-sized operations were carried on in that county with varying degrees of success.[53] In other areas kaolin, slate, marble, lime, lumber, and turpentine industries were developing rapidly.

A stone quarry on the Connelly plantation near Augusta produced burr millstones which were cut and processed by the LaFayette Burr-Millstone Company in Savannah.[54] "If the pair sent [to us] is a fair sample of the Georgia Burr Block, we do not see the necessity of another importation of the article from France," wrote a Richmond firm in 1850.[55] An Augusta firm, capitalized at $80,000, began working the valuable kaolin beds near that city. Already Augusta was using local kaolin pipes in its new water works and drainage system.[56]

Georgia clay was sent to New York for analysis in 1846 and was adjudged comparable to the Devonshire clay used in the popular Staffordshire designs. "I consider the discovery of this China clay to be very important," commented the editor of the *Southern Cultivator*.[57] Dennis Redmond, experimenting with cheap building materials, constructed a Negro house with "clay mortar" or tapped earth, to which he had added small portions of chopped straw. The walls, which were a foot thick and coated on the outside with cement, were further protected from rain by a two-foot roof extension. To prevent the walls from absorbing moisture a cement foundation extended a foot above the ground.[58] The floor was made of pounded clay topped with cement.

In order to utilize the products of lime kilns then being developed in Georgia, it was suggested by some that more concrete be used in the construction of farm buildings.[59] Slate from the newly opened Blance quarry in Polk County was finding wide markets in the 1850's.[60] Operators of the Van Wert quarry in the same locality offered slate "laid on the roof at any point in Georgia, Alabama, and Tennessee at an expense not exceeding the cost of tin roofing."[61] Worm rail fences were being replaced on a few plantations by strands of wire, but since the wire was not a Southern product, this type of fence did not become generally popular.[62] To supply fencing material in the older areas, the growing of cedar in abandoned cotton fields was advocated. Some suggested that the shortage of fencing wood might be alleviated by revising the existing stock laws so that fences would not be required.[63] There was no dearth of timber in the Pine Barrens, however. At Savannah the lumber export business increased each year between 1840 and 1845. The Georgia pine in this period enjoyed undisputed pre-eminence over all other American pines.[64] The turpentine industry also flourished along the coast, where distillers in 1847 obtained legislation for the protection of their industry.[65]

The value of cottonseed both as a fertilizer and as food for stock came to be recognized in the 1850's.[66] In announcing the establishment of a cottonseed-oil mill at New Orleans in 1854, the *Southern Cultivator* estimated that thirty-eight million dollars annually was lost to planters by their failure to utilize their cottonseed. The *New Orleans Delta* stated that oil from cottonseed could be used for manufacturing soap, candles, paint, and cosmetics. While oleomargarine was yet unknown, it is significant to note that cottonseed oil was suggested as a substitute for butter. The *Baltimore Daily Times* prophesied that this oil would one day become a source of great revenue for the South.[67]

Interest in the artificial fish pond became popular in the 1850's, although development of lakes and ponds was not widespread. John C. Carmichael maintained four large lakes on his place at Greensboro, and in 1859 announced that he expected to produce that year a total of ten thousand trout, averaging ten inches in length. In his "Loch Lomond Pond" were such varieties of fish as bream, white perch, red horse, mullet, and suckers. William Gesner,[68] a Columbus druggist and erstwhile resident of Milledgeville, was perhaps the outstanding authority in Georgia on pisciculture. He described with minute fidelity the life, migration, and spawning habits of native Georgia fish. In 1858 he was engaged by private citizens in Alabama to stock streams in that state with white shad trapped in the Oconee at Milledgeville.[69]

One of the more significant accomplishments in the quest for new farm crops came in 1855 when sorghum cane was introduced by Dennis Redmond of Augusta, one of the editors of the *Southern Cultivator*. While the production and manufacture of sugar had long been a minor enterprise of coastal planters, the practical supply of food with a high carbohydrate content remained a problem in the upcountry where cane sugar and molasses were relatively expensive.[70] Indicative of the demand for such a food was the prominence given to experiments in the manufacture of sugar from Indian corn and other products.[71] While a plant known as "ribbon cane" could be grown in the southern part of the state and made into syrup, it did not thrive north of the fall line.[72]

In 1854 sorghum seed were brought to America directly from France, where the crop had been introduced by the French consul in China three years earlier.[73] Dennis Redmond acquired a few ounces of the seed which an enterprising firm in Boston had just imported.[74] He planted six or eight hills in his garden early in the spring of 1855 and at the same time distributed some of the seed to Richard Peters, Robert Battey, Robert Nelson, Jarvis Van

Buren, and John Bonner, all living in different parts of the state.[75] The new plant was known variously as Chinese sugar cane, *holcus saccharatus,* and *sorgo sucre.* "We are not aware that it has been tried in this country," Redmond announced, "but it would doubtless succeed well in the Southern States and deserves to be tested, if only for fodder."[76] In the following year the editor of the *American Agriculturist* made a similar distributon of seed in several regions of the Upper South. In the meantime, however, Redmond had sent some of the seed as far west as Texas and announced that he would make no further free distribution.[77]

The new crop was heralded with more than usual enthusiasm. The claim that it would give planters in the interior both syrup and sugar was somewhat exaggerated. A more modest claim was that it would furnish excellent feed for livestock, and two crops could be grown in one season from a single planting. It was also said that the seed alone would produce fifty bushels of good meal to the acre, and the bagasse might have commercial uses, such as in papermaking.[78]

Perhaps the first reported experiment in America on syrup and sugar manufacture from sorghum took place in North Georgia in 1856. The wife of Richard Peters' overseer succeeded in boiling some of the juice into syrup, and Peters announced in September that he would install equipment and process the syrup in commercial quantities. He improvised the necessary equipment and engaged Robert Battey of Rome to help him work out technical details.[79]

Battey was at that time on vacation from his study of chemistry at Philadelphia. He conducted the experiment with all the care and precision of an important scientific venture and reported the results in the same style. The stalks were weighed before grinding and the juice was weighed before and after boiling. The specific gravity of the raw juice was also ascertained. It was then boiled with lime, and other clarifiers were tested. In translating a paper by a French scientist, Battey discovered that Georgia-grown cane yielded a larger quantity of syrup than that produced in France.[80] He succeeded in making sugar from the juice but he was not optimistic over such possibilities. The cost of growing and processing the syrup he estimated at fifteen cents a gallon. From the results of these experiments Dennis Redmond prepared a pamphlet on syrup making from the new cane, to answer inquiries which were pouring in from enthusiasts throughout the United States.[81]

Knowledge of the approved method of converting the juice to

syrup was not generally diffused to all farmers interested in the new enterprise, however. The result was an inevitable glutting of the market with thousands of gallons of overcooked, unpalatable syrup, or else a thin product resulting from insufficient boiling and which quickly soured. The over-feeding of the green cane to cattle and horses resulted in some fatalities among livestock.[82] These circumstances caused many to consider the experiment a failure.[83] David Dickson expressed his belief that it was "not worth raising for any other stock except hogs."[84] Some referred to the cane as Chinese broom corn, and it came to be confused with such African grain sorghums as the kafir corn. Thus it was ridiculed as another humbug.[85]

The campaign of denunciation was encouraged by growers of sugar cane in the coastal regions, who recognized sorghum as a potential competitor to their own product. A rumor alleging that sorghum syrup had caused the death of fifteen Negroes on the plantation of Benjamin F. Adams of Lee County was promptly denied by the owner. A Mississippi planter denounced the cane because it could be grown in the North and would thus infringe upon the market for Southern sugar and molasses. "I cannot see why we should be seeking out information for the benefit of the North," he wrote. "Could I control the products of the South, I would put these Negro stealers of the North on this Chinese Broom Corn molasses and scratch them with hemp shirts . . . ."[86]

Indeed the early speculations on the economic possibilities of sorghum indicated that it might become a greater boon to the Northern farmer than to the Southerner. A committee of the United States Agricultural Society noted that the soil and geographic range for sorghum corresponded with that of Indian corn and that it required approximately the same culture.[87] A Pennsylvania paper in 1858 predicted that "the Middle States of this Union are to become, ere long, the greatest sugar-producing country in the world."[88] A Chinese Sugar Cane Convention was held at Springfield, Illinois, in 1858, at which various cane products were exhibited, including sugar, molasses, and rum.[89] An Ohio farmer devised a shallow evaporator which replaced the clumsy iron kettles, and which afterward remained the standard type of equipment for processing.[90]

It was soon discovered that, while the yield of juice was always about fifty per cent of the total weight of the cane, the yield of syrup varied from ten to twenty per cent, depending upon the latitude in which the cane was grown. Happily for Southerners the conditions prevailing in the latitude of the Georgia Piedmont produced the highest yields. There was also evidence that Southern-grown cane produced a better quality of syrup.[91] While many white men never developed a taste for the product, it became a standard food in the diet of Negroes. The naval blockade of the South during the Civil War and the impoverished conditions of that section following the return of peace helped to place the industry on a permanent footing. Sherman's bummers reported large supplies of sorghum syrup "at nearly every plantation" on the march to the sea and his soldiers developed a liking for it. During the war it was known as Confederate syrup.[92]

The most effective method of cultivating sorghum cane was described by Robert Nelson. The cultural methods which he suggested as early as 1857 were still practiced in Georgia a century later. Among other things he cautioned against pulling off the blades for use as fodder and pinching off the seed before the cane was fully ripe. Anything which caused the cane to throw out suckers injured the quality of the juice, he said. For preserving the canes until fall when other canes would be coming to the mill, he advised stacking in the barn with alternate layers crossing at right angles, to a height of twelve inches and then covering with bagasse to prevent drying. This curing process would produce a better quality and a higher yield of syrup. He described a silage with sorghum as a base which would be available in dry seasons when the corn crop failed.[93]

Another product from China made its appearance in Georgia in 1856—the "Chinese prolific pea," having made its way to the South from the West Coast. It was grown by many farmers in the vicinity of Augusta, where Robert Nelson discovered that it was a member of the bean family and reported that it was a poor substitute for the native pea. Its soil-building properties being unrecognized at the time, it ceased to attract attention until after the Civil War when it again came into prominence as the "Southern Relief Pea." Catalogues named it *Soja hispida,* and it became the soy bean of a later era.[94]

The native or "Southern" pea underwent rediscovery in Georgia around 1850. This pea, which was little known outside the South, had been cultivated by the Southern Indians and was often called the Indian pea, or the field pea, to distinguish it from the English garden variety. Prior to 1850 its value as a soil builder was not fully recognized, but it rapidly came into use as a food for both man and beast. Daniel Lee called it "the clover

DIVERSIFICATION AND ECONOMIC INDEPENDENCE   87

of the South."[95] General Robert E. Lee is said to have claimed that the Confederate soldier's remarkable capacity for survival on limited rations was due to his use of the pea as a food.[96]

Considerable progress in diversification was recorded in Georgia in the decade beginning in 1840. Census data show that the per capita figure for livestock doubled in the ten-year period, with cattle and sheep doubling in number. Of the various food crops grown, only wheat showed a decline, apparently the result of a heavy sleet storm throughout the northern part of the state in April 1849, the crop year on which the 1850 figures were based. Gains made in the production of oats, barley, and rye more than offset the loss in wheat, however. Decided gains were recorded for sugar, tobacco, and hay. The trend throughout the decade was definitely away from cotton and in the direction of grains and livestock. The record book of George J. Kollock, a planter near Savannah, shows the products shipped from his plantation throughout a number of years. Until 1844 cotton was the chief product. After that year appeared frequent notations of corn, rice, cattle, and wool. Minor crops consisted of potatoes, peas, and sugar cane, all for use on the plantation. Occasionally potatoes were sent to the market in Savannah, as were a few dozen oranges.[98]

The plantation of James Rucker in Elbert County comprised more than thirteen thousand acres in 1860. He had livestock of all kinds and produced immense crops of provisions. Negro artisans made meal, flour, lumber, plows, harness, and shoes at the plantation mills and shops. They did all of the primary processing and much of the finished manufacturing required of the plantation community of more than two hundred people.[99] The well-known plantations of Farish Carter in Baldwin County, David Dickson, and William Terrell in Hancock were noted for the balanced economy which their owners practiced.[100] William Terrell's estate in 1855 included 640 head of cattle, 225 hogs, 6,000 bushels of corn, and various quantities of fodder, wheat, oats, and peas.[101] Emily Burke visited a plantation near Waynesboro which covered forty-nine square miles on which she reported finding both social and economic self-sufficiency. Negro women plowed alongside the men, and the latter shared the society of quilting parties and other feminine social-economic functions. The principal overseer was a Negro slave, trusted by the master and respected by his charges. Holidays were frequent and slaves were permitted to cultivate patches of their own and sell their products at market prices.[102]

Solon Robinson found ex-Governor George R. Gilmer retired to his farm near Lexington in 1851. Like a Roman ex-consul, he was growing cabbages and enjoying the serenity of rural retirement, yet lamenting the fact that cotton absorbed too much of the planter's time. He had a large library of valuable books and he was making a collection of Georgia minerals. In the pursuit of the agrarian life he was said never to have grown a bale of cotton for the market.[103] The following description of Pleasant Valley Plantation, the home of Dimos Ponce in Hancock County, closely approaches the agrarian ideal of an independent economic unit.

[The plantation contained] 800 to 1000 acres, with a two-story house on a stone basement, with two wings for a library and Greenhouse, and other needful outbuildings, a large Barn and Gin house, together with a stock of mules, cattle, sheep, and hogs; also, wagons, carts, carriage, Buggy, two cotton Gins, wheat Threasher [sic], Corn sheller, Corn, Fodder, etc. . . . The place is well stocked with fruit trees of the most approved varieties—Pears, Apples, Peaches, Nectarines, Plums, Apricots, Cherries, and a small grapery. The plantation is well-watered with springs and rivulets. There is a well of good water in the yard and a fine spring at a convenient distance from the house.[104]

On thousands of farms and small plantations a subsistence economy had long been practiced, and the fall in cotton prices in the early 1840's perhaps wrought little change in their customary planting procedures. Alexander Thomas, a bachelor living in Meriwether County in 1852, did his own cooking and made eleven bales of cotton, forty-five barrels of corn, twenty-five bushels of potatoes, five thousand bundles of oats, and some wheat, fodder, and vegetables.[105] George Gilreath moved to Cass County in 1838 without money or property. In 1860 he possessed 2,100 acres of land on Two Run Creek, a comfortable residence and outbuildings, and ample livestock. "With a small force of Negroes and eleven children of various ages, the average yearly productions on his land amounted to 2,000 bushels of corn, 1,000 bushels of wheat, 1,500 bushels of oats, and 18,000 pounds of pork, in addition to mules, horses, wool, mutton, beef, and hay." Charles W. Howard attributed Gilreath's success to his policy of planting no cotton. "Here we have an instance of a man going into the woods, without money or negroes, taking hold of an unpromising piece of land, turning it into a valuable farm, raising a numerous and respectable family, and at middle age finding himself surrounded by every comfort of life," wrote Howard.[106]

Interest in diversified economy was never confined entirely to those who might profit directly from such practices. Garland D. Harmon, an overseer, was one of the South's better-known spokesmen for economic independence.[107] His numerous contributions to agricultural journals before 1861 reveal an interesting transition in his thinking, moving from a practical allegiance to diversified economy, to total economic independence, and finally to Southern nationalism and political independence. "Time has demonstrated the fact, that the doctrine which you and I . . . advocated so long and earnestly of making the South independent of the North and Northwest, was correct," he wrote to Charles W. Howard in 1863, a few days after receiving a surgeon's discharge from the Confederate Army. "Had these doctrines been heeded then . . . there would be no necessity now for legislative action and executive appeals on the subject of subsistence."[108] While Harmon never owned a farm, his nationalism was unmistakably a product of the soil which he loved and whose fertility he husbanded with fanatical devotion in the service of others.

The average Georgia farmer in 1850 was neither a landless hireling nor the traditional planter. He was a practical farmer who diversified his activities when he thought cotton likely to be unprofitable. As profits from cotton assumed increased promise, as it did after 1850, he moved cautiously in the direction of the one-crop system, but it was a modified system, wholly unlike that which prevailed earlier. He was careful to grow enough food for his family and livestock, the corn crib being his insurance against the hazards of low cotton prices. These were simple principles which he learned by long and sometimes painful experience. While realistic in the approach to his own problems, he was no student of the deeper problems which faced Southern agriculture. Conservative and responsible, he relied on hard work and the intensification of his customary efforts to solve the problems of a diminishing income.

A close relationship existed between diversification in agriculture and the development of industrial enterprises in the decade and a half before 1860. Each was a manifestation of a rampant spirit of economic nationalism. Before 1840 manufacturing in Georgia was confined largely to the plantation shops and the farmhouses, and it was closely integrated with farming activities. In that period Southerners had despaired of ever becoming a manufacturing people. During the agricultural depression of 1828 the editor of the *Southern Agriculturist* recommended to Southerners "not to meddle with manufacturing . . . even for the supply of their own plantations," and he cited the absence of satisfactory labor. "Shall we be more successful in making overseers of cotton weaving and spinning, than we have been in making . . . overseers of our plantations?" he asked.[109] There were few Southerners indeed so optimistic as to believe that the South could ever do more than perform the primary manufacturing processes. Thomas Spalding was among those who believed it advisable to appropriate a hundred thousand dollars to subsidize the manufacture of coarse cotton clothing.[110]

Home and farm manufacturing received an impetus from the low cotton prices of the early forties. Such part-time plantation manufacturing would profitably utilize unskilled labor during slack seasons. William Terrell in 1852 suggested to the Cotton Planter's Convention a premium for simple spinning machines which might be installed on plantations. He cited experiments proving that Negroes could become expert operatives of such equipment.[111]

John Cunningham of Greensboro, an erstwhile cooper by trade, described himself in 1850 as "a broken down planter and farmer" who had been compelled to leave farming. He studied the writings of William Gregg, the South Carolina industrialist, and came to the conclusion that no exclusively agricultural country could ever be consistently prosperous. With James L. Coleman he established a large flour mill on the Augusta canal and offered premiums for the stimulation of wheat growing in that vicinity.[112] "If I only had the pen and the head . . . as I have the heart, I would [contribute something to] make Georgia . . . the pride of the South," he said.[113]

Manufacturing in Georgia in the 1850's was designed primarily as a complement to agriculture—a kind of agrarian industrialism. Manufacturing was justified only if it would enhance agriculture by supplying a greater portion of the planter's needs and so enable the South to achieve a greater degree of self-sufficiency. Thus the older concept of crop diversification within the individual plantation evolved into a move for economic independence within the community and finally to the broader concept of economic nationalism for the entire Southern region. The agrarian-industrial leaders were visualizing an economic system balanced in all its parts. Census data show that home manufacturing had almost disappeared in the Georgia Piedmont by 1860, except among the lower social and economic groups residing largely in the more backward regions of North Georgia and in the Pine Barrens, where the practice survived until the next century. This work was done almost entirely by females and was limited usually to clothing.[114]

On the eve of the Civil War a correspondent to the *Southern Cultivator* noted the decline of home manufacturing and lamented the trend, which he falsely interpreted as regression to economic dependence upon the North. "We are the veriest dependents on earth, and no wonder we are vassals," he chided. "None is so great a slave as him who has not the spirit to be free . . . . I have seen a factory with fifty to one hundred sewing machines, driven by steam power and attended by Yankee galls [*sic*], making up Negro clothing."[115] What he failed to realize was that manufacturing enterprises had merely been transferred from the home to the local factory. By 1860 factories were found in almost every town throughout the Piedmont, and in many rural villages.

It was in the last decade and a half before 1860 that Georgia came to be referred to generally as the Empire State of the South. "The path of empire is open to us," said the *Soil of the South* in 1851. "The soil is here—let us cultivate it and retain it; the raw material is here—let us manufacture it; the water-falls are here—let us dam them and turn their now wasted energies upon our wheels; the ores are here—let us search them out, and turn them to service."[116]

An example of the type of agrarian industrialist in Georgia in this period was Mark A. Cooper of Cartersville. The Cooper Iron Works embraced a mining field of twelve thousand acres and had a capital investment of half a million dollars. The establishment boasted a nail factory, coal and iron mines, and also a flour mill with a capacity of three hundred barrels a day.[117] Cooper was educated in Georgia and South Carolina and was admitted to the bar at Eatonton in 1821. In 1833 he organized a company which built on the Little River near Eatonton one of the first cotton factories in Middle Georgia. He later engaged in banking at Columbus, and was elected to the 26th Congress as a Whig. Defeated for the governorship by George W. Crawford by a very narrow margin in 1843, he retired from political life, becoming one of the founders and an official of the Southern Central Agricultural Society.[118] He consistently advocated diversification and the development of Georgia's latent resources. Like other Southerners such as William Gregg, Daniel Pratt, Edmund Ruffin, and Joseph Reid Anderson, he became a vigorous advocate of Southern economic independence. He believed that Georgia's minerals could be developed to possess an industrial value equal to the value of its cotton.[119] "Georgia now imports millions of Dollars worth of Iron and Iron products; also products of the Farm, Orchard, Garden, and Dairy . . . [and] Cotton pays for it," he said. "Without this policy what has she in prospect but worn out soil and decaying institutions? Where are her means of defense? . . . What nation or people ever did otherwise than submit, who are fed and clothed by others, and who are even dependent on their oppressors for their axes and hoes, their plowshares and their prunning hooks, as well as their swords, their guns and their powder."[120]

In 1860 there were more than eighteen hundred manufacturing establishments in Georgia employing a total of eleven thousand laborers. These establishments, making the less refined products, were closely related to agriculture. The products having the greatest annual value were flour and meal, being valued at a million and a half dollars in 1859; and the business of making them was also the most highly competitive. The largest capital investment was in the cotton goods industry, while sawmilling required the highest labor cost.[121] By 1850, railroad transportation had begun to transform certain rural villages in North Georgia into budding centers of industrial development. This interest was reflected in the activities of various agricultural societies in the region. The Etowah Agricultural and Mechanical Association had perhaps the largest number of manufactured articles on any premium list to be found in Georgia. Included was a great variety of agricultural tools and equipment, leather goods, furniture, flour, shingles, ropes, and clothing.[122]

"The Empire State" by 1860 had won a regional reputation for agricultural diversity and industrial development. "Georgia has set an example of wisdom, and very soon she will possess within herself so completely all the elements of empire that she might forever be separated from the surrounding world and yet flourish," wrote a citizen of Louisiana. "Upon her hilltops begin to smoke the wealth-achieving furnace; the buzz of the cotton spindle mingles with the whisperings of her clear blue streams; the 'iron horse' is far and wide circulating her products . . ."[123]

VII

Soil Conservation and Southern Nationalism

THE OLD BAR-SHARE plow had largely gone out of use in Georgia by 1840, being replaced by the bull-tongue, scooter, and various shovel plows.[1] These plows generally were triangular in shape, the cutting point being a forty-five- or fifty-degree angle. The bull-tongue, used for breaking new land, was less of a triangle than the others, having somewhat convex lines toward the point and a surface curvature resembling the tongue of the animal for which it was named. The scooter, a variety of the bull-tongue, received its name from the ease and efficiency with which it scratched the surface of the soil. The shovel plowshare, shaped like the letter V, was ridged in the middle with wings sloping toward the rear. It was drawn foremost through the ground. Its detractors characterized it as being like a cat dragged by the tail, and its ability to stir the soil was said to be only a little more effective. Perhaps seven-eighths of all plows in Georgia in 1840 were of this class.[2]

After 1840 a few improved plows of Northern manufacture came into use, and a number of superior plows were made in local blacksmith shops. Among the latter were the Dagon plow, a crude, one-horse turning plow with a wooden mouldboard and a wrought-iron share, adapted to light soils; the nullifier, a broad-pointed plow designed to turn the soil half over, used for breaking stiff stubble, cutting deep-rooted vines, and for bedding cotton land; and the single coulter, which usually had one wing attached to the side and required special stocks.[3]

Secondary plows, such as the harrow and the roller, were practically unknown in Middle Georgia before 1840. It was not until about the beginning of the Civil War that there came into use a cotton-planting device which would open a furrow, sow, and cover the seed in one operation instead of three. "We are in great need of all sorts of farming tools," wrote a Walker County farmer in 1842. "If some good mechanic would come to this country and make all such things, and keep a tool warehouse, I think he might make a fortune."[4] The great variation in soil texture and topographic features throughout different regions of the state precluded the standardization of plows and minimized the value of experimental data accumulated under such a wide range of conditions.

Travelers and settlers from Northern states were uniformly impressed by the simplicity and inefficiency of agricultural implements in use. A "Georgia plough" was described in 1845 as "a rude misshappen uncouth thing," the remnant of ignorance and primitive conditions.[5] "I wish you could see a southern plow. . . . It would be a great curiosity to a New York farmer," wrote Jarvis Van Buren to a New York journal in 1851. "I am sure he could not tell for what use it was intended, or to what nation of people it belonged."[6] Paul Davidson, who had recently settled on a farm at Raytown, wrote to the *Albany Cultivator* praising the mild and salubrious climate and the fertility of Middle Georgia, but he condemned "the system of skimming with a set of queer shaped irons, misnamed plows, which loosen about as much of the surface as the summer's rains can easily carry down the steep hillsides . . . . [leaving] a mournful spectacle of galls and gullies . . . ." While he lamented the loss of fertile top soil, he recommended the section for the low price at which land was selling and the ease with which it could be improved.[7] A Northern agricultural writer, surprised to find that an acre of Clarke County land was unable to produce more than five hundred pounds of small grain and straw, thought one reason was "for want of better plows, [so] that a few days of sun exhausts all the moisture, and leaves the tender plants to struggle for life in a bed of dust . . . ."[8] He quoted an Athenian who insisted that cast-iron plows were undesirable because they buried the fertile top soil so deeply that nothing would grow afterwards. He believed that two inches was deep enough to plow and that the old fashioned shovel plow was the best implement ever invented for the purpose.[9]

Agricultural reformers, on the contrary, recognized that defective plowing with poor tools was the chief cause of soil waste. "We must have another race of blacksmiths as well as improved implements," wrote Jethro Jones, "for the work of their hands [has] aided in desolating one of the fairest portions of creation." Such, said he, had been the fate of Clarke County and ". . . . nothing is left . . . save the history of her old red hills."[10] A Butts County farmer described plowing with the scooter and shovel as cutting into the ground a sharp trench which was only reopened in subsequent tillage, leaving a hard unbroken ridge on either side.[11] Another noted the tendency to spread plowing as thinly as possible when the highest perfection could have been obtained with intensive cultivation. He described the prevailing practice as "scratching the surface of the earth indiscriminately from two to four inches deep . . . the furrow representing a triangle with its vertex turned downward." On lands of considerable declivity, he said, such plowing caused washing to start in the gutters formed in the hard subsoil. He suggested the use of twice the number of horses to the plow, and some kind of improvement over the bull-tongue share.[12]

Daniel Lee, who thought the absence of labor-saving machinery the most striking feature of Southern agriculture, was a strong advocate not only of improved plows but also more farm machinery of all types. He advised the use of steam power for stationary machinery so as to render horses available for better plowing.[13] He suggested the arrangement of five or six bull-tongue plows in a gang, like teeth in a cultivator, to be drawn by four mules, thus to minimize human labor.[14] By adopting Northern gang plows, he thought that nine-tenths of all cotton and corn fields in Georgia could be plowed at two-thirds the normal cost. "Why not have nine single mules, (and as many hands to drive them) to pull nine separate harrow-teeth through the surface soil?" he argued. "Why not keep six little wagons, (instead of one,) with six drivers and six mules and haul six little loads of cotton or wood to market, where you now haul a single one?"

Admitting that Southern machinery was "killing to horse flesh," the commentator "Broomsedge" called Lee's gang plow a Yankee horse-power contraption. "Think of them four plows of his in a gang held by a boy and drawn by three mules," said he. "Wouldn't they go it in a stumpy new ground . . . . Why Dr., before 9 A.M., the nigger would be knocked into fits, and the mules gone to grass, and the plows out of gearing."[15] The adaptablility of cotton and corn to simple tools and the widespread use of Negro labor contributed much to the backwardness of field culture in the South. Also the tradition of planting cotton in newly cleared ground with rough features required simple plows operated by single mules. This situation in turn conditioned planters to demanding the cheapest plow they could buy. "[The mechanics] know by sad experience that they cannot afford to put up a first rate plow for the prices the planters are willing to pay," said a farmer in 1858.[16]

Mechanical skill was not of a high order in the Cotton Belt. A few intelligent Negroes had been trained in such skills as blacksmithing, but a relatively small number of foreign-born white artisans were the chief source of recruits for these occupations. The antipathy of native white men toward a trade appeared to be least in the urban centers. Some believed that the mechanical trades could be made more respectable by confining Negroes strictly to field work.[17]

Perhaps the most ideal tool developed for cotton culture in the ante-bellum period was that patented in 1855 by George W. N. Yost, a Mississippi overseer.[18] A combination plow and scraper, it could at a single round pulverize the middle of the row and yet scrape lightly the area nearest the plants, thus reducing to a minimum subsequent work of hoeing by hand. Manufactured in Pittsburgh, it sold at the relatively high price of twelve dollars.[19] "If any man on earth is benefitted by [it] it is the poor man who toils alone in the field," said a farmer who recommended it to Georgia planters.[20] This tool remained part of the standard equipment of cotton farmers for more than a century.

David Dickson in 1860 was advertising a turning plow with a sweep which was being manufactured by experienced Negro smiths under his supervision. This implement, a modification of Yost's tool, was used for the same purpose. It found a ready sale among farmers in Georgia and adjoining states. The earlier type of sweep was a semicircular piece of wood lying flat on the ground with a cutting edge of steel slightly lowered into the ground.[21] The Dickson sweep, sometimes made by welding two narrow wings to a scooter, appears to be the real progenitor of the wing scrape of a later era. While the edges dulled easily and the rough surface clogged in sticky soils, it was an ingenious contribution to the limited variety of Southern plows.[22]

There were evidences of other improvements in Southern plows by the end of the period. Throughout the 1860 decade there was a growing list of patents issued to Southerners for farm machinery.[23] "The mechanics of the South are beginning to stand upon their true elevation," said Garland D. Harmon. "They can, and will, in a few years, supply the planting States with all their implements of husbandry."[24] In 1855 a Georgia farmer expressed his opinion that the amount of labor performed by a single hand in Georgia had almost doubled since the Colonial period.[25]

The increasing use of better plows resulted in improved methods of soil culture. The practice of laying off rows along the natural contours of the land was first introduced into Virginia from Europe by Thomas Mann Randolph around the end of the eighteenth century.[26] Known as horizontal plowing, the system did not come into use in Georgia until after 1830 when it was advocated to limit erosion resulting from heavy falls of rain. However, contour plowing was not practiced universally in Georgia at any time during the ante-bellum period. David P. Hillhouse expressed the prevailing sentiment when he said: "Deep plowing is very necessary, but I do not conceive serpentine and zig-zag beds and water furrows to be required."[27] David W. Lewis described the tillage that prevailed in Hancock County as late as 1849 when he said that "the rows . . . went straight across the field, up and down hill."[28]

Horizontal plowing alone worked very well on gently rolling land and under conditions of moderate rainfall, but during very heavy rains, particularly on steep hillsides, the water broke over the sides of the rows and caused gullies. To mitigate this evil horizontal plowing was combined with hillside ditching, the object of the latter being to provide drainways at intervals in the field into which water from the rows could flow. The hillside ditch was a transitional stage in the development of the terrace of a later era.[29]

John Farror claimed to have introduced hillside ditching into Middle Georgia around 1833, having brought the idea from Virginia when he moved to Putnam County. However, Richard Shipp Hardwick, of Hancock County, appears to have begun the practice in 1832.[30] It is certain that Hardwick did more than any other man in Georgia to assemble and disseminate information on the method, and to popularize it. He published several articles in agricultural journals, explaining how to operate a grade level in running the ditches. His outstanding essay, expressed in simple language, was published first in the *Farmers' Register* in 1842. His article appeared later in the *Southern Cultivator,* and subsequently was republished in every leading agricultural journal in the South.[31] The *Southern Cultivator* and the *American Cotton Planter* ran a second series of Hardwick's articles in 1856 at the request of their readers who were beginning to take a renewed interest in soil conservation.[32]

The experience of a planter at Covington is typical of those who lacked experience in soil conservation. This planter whose "cow hide farm" had been cultivated for fifteen years on the "down-hill fashion," described the soil as "about as thick over the clay as the hide of an ox." Although a reader of agricultural journals, he was unable to increase productivity by deep plowing and manuring. About 1844 he began to plow in contours, but during heavy rains water broke over the rows to cause gullies. He then pursued Hardwick's plan for constructing hillside ditches but gave them less fall than Hardwick had recommended. He stated the results of the next heavy rain as follows:

> Many of the ditches were filled with sand, and others demolished; some gullies were formed and others deepened; and much of my land, that I had title "to have and to hold in fee simple forever," nullified my deed, and started down the river toward Macon .... When I have seen not only the soil but the manure that I have toiled so long to get on my fields, soon washed away, I have felt dispirited and sad; and visions ... of the fruitfulness of Alabama, Mississippi, and even Texas, infested my mind. But the ties of nativity, friendship, kindred and comfort, have thus far bound me to the soil of my ancestry.[33]

With a few adjustments he succeeded in making the system work, reaching the conclusion that "emigration to a wilderness country with its thousand attendant ills" was no longer necessary. He summed up his philosophy by saying: "I know that ditching *per se* cannot enrich land, any more than the President *per se* can annex Texas. But let one be backed by manure and the other by the Senate, and both objects will be accomplished...."

The system spread from Georgia into the Southwest. Hardwick was lauded throughout the Lower South for this contribution to the agriculture of the region. A Tennessean in 1856 affirmed that he had become "well nigh disgusted" with his success as a practical farmer until he read Hardwick's articles.[34] A Houston County (Georgia) farmer suggested a monument to him as a public benefactor. "My farm is worth ... treble what it was, and I am debtor to Mr. Hardwick for it," he said. Andrew J. Lane listed Hardwick's contribution as first in the factors responsible for the rejuvenation of agriculture in Hancock County.[35]

The techniques of horizontal plowing and the construction of ditches, while relatively simple to some, were not easy for many farmers to master. Hardwick recommended that the ditches be spaced from seventy-five to one hundred and fifty yards apart, depending on the fall of the land, but thirty or forty feet apart on very steep hillsides.[36] The grade of the ditches was regulated by the ability of the soil to absorb water, clay lands requiring a greater fall than sandy lands. A common rafter level could be used in running the grades, in lieu of which a wooden frame was constructed in the shape of an isosceles triangle. The latter had a base of twelve feet and the two sides extended to form legs below this base. From the apex a plumb bob was suspended by a small cord. A cross-bar fixed at right angles to the plumb line, with center markings, completed the instrument.

The operation was begun at the highest point of the field and about the middle of the intended ditch to be constructed, so as to cause the water to flow each way from the center. After the ditch was laid out it was cut about a foot deep and from twelve to twenty inches wide. Guide furrows were then laid off on a true level, striking the ditches and crossing them where necessary. The space between the guide furrows was next laid off in parallel rows, with short ones where they were required. Short rows were made necessary because the guide rows sometimes ran close together as a result of an abrupt decline in the land. Thus a field of irregular topographical features was theoretically transformed into a single inclined plane which conveyed the water off gently.[37]

Many beginners, while laying off and grading the ditches properly, made the error of running the rows parallel to the ditches, failing to see that the natural undulations soon varied the grade and changed the direction of the water at different points on the row. Hence the water was thrown into great bodies at one point in the field and soon increased, causing gushing overflows.[38] This was the common error which caused the entire system to be criticised as wasting too much land and necessitating tedious short rows on rugged terrain.[39] Many laid off the ditches by guess, but even when done by the aid of an instrument, the task was often performed poorly. The ditches might be constructed with or without horizontal rows. To obtain best results the surveying was done on land leveled and harrowed as in preparation for planting wheat, but this precaution was not always taken.[40]

With such confusion and lack of mastery of the technique, it is little wonder that many Georgians adopted the system slowly and with misgivings. An observer in 1851 noted an alarming absence of hillside ditching in Clarke County, and was told by a prominent Athens planter that the method was of no use in holding the rows "because the rain falls in such torrents, it fills up or sweeps them all away."[41] Judge Augustin S. Clayton, while holding court at Watkinsville, was inclined to sustain a young lawyer who defined land as movable property, for the sake of the moral lesson for those who refused to conserve their soil.[42] An editorial in the *Federal Union* in 1850 expressed both surprise and enthusiasm at finding a planter in Twiggs County who, by horizontal plowing and hillside ditching, had ended the washing away of his soil.[43]

Only in Hancock County, under the stimulus of Hardwick's own plantation which served as a convenient model, was the system gaining anything like widespread practice before the Civil War. There Daniel Lee noted the enthusiasm with which farmers were inaugurating the system. He also noted the improved and settled appearance of the homesteads. He observed young Thomas Grimes, who had just come into possession of a 4,500-acre plantation, erecting a new residence. Lee found Grimes in the fields working with his Negroes, superintending the construction of hillside ditches. "Others may skin the earth and roam like savages from one wild to another," he mused, "but for me and my offspring, here is our home." Lee suggested in 1857 that if intelligent young men should qualify themselves as "Engineers of hillside ditching" and travel through the country they would find remunerative employment.[44] An effort was made to secure by public subscription the services of an expert who could tour the state and explain the system to farmers, but the movement collapsed for lack of proper backing.[45]

Garland D. Harmon was one of the South's more persistent advocates of hillside ditching, horizontal culture, and improved plows. The last he insisted should be made by Southern mechanics. His early correspondence illustrates a fanatical devotion to Southern soil, then a transition to economic independence, and finally to Southern nationalism. He early mastered Hardwick's system of horizontal culture and applied it to plantations that came under his supervision, vigorously advocating it throughout the various communities in which he found employment. Between 1849 and the close of the Civil War, he contributed nearly one hundred articles to various agricultural journals. While he discussed every phase of plantation economy, his great passion was the improvement of the soil of his native Southland. Because he was an exceptionally intelligent and responsible member of the lowly class of overseers, his career deserves more than passing notice.[46]

Born in Georgia in 1823, Harmon began the career of overseer on the 2,000-acre plantation of Thomas H. Sparks near Cedartown, in 1847. Two years earlier he had married Emily Edge, a student in Hearn Academy in Cedar Valley, where her father owned 240 acres of land. At that time Harmon lived on a five-acre lot which he owned within a stone's throw of the Cave Spring Academy, where Cedar Valley joins Vann's Valley in Floyd County. In 1847 he sold the lot for two hundred dollars, stipulating that the purchaser never erect a gaming house or deal in intoxicating liquors, thus indicating the decorous conduct which he manifested throughout his life.[47] As one of his father-in-law's five heirs, he sold for seven hundred dollars in 1852 his part of John Edge's estate, together with a half acre of land "which he owned in his own right."[48] Thus at the age of twenty-nine his short and uneventful career as a landowner had ended for all time, and his career as an overseer on the land of others was well begun.

Harmon's status as a landless farmer by no means diminished his agrarian philosophy and his interest in the conservation of Southern soil. Before 1852 he wrote many letters to the *Rome Courier,* in which he recommended to Floyd planters new tools and new methods of plowing, and criticized those who resisted all innovations for the improvement of agriculture. He took a prominent part in the Floyd County Agricultural Society, serving on its important committees with men who bore titles of esquire, judge and colonel. His exhibit of tools won a premium at the fair of 1852.[49]

His writings revealed at least three powerful aversions: "old fogyism," which ridiculed book farming and agricultural experimentation; "down hill farmers," who left great acres of gullied hillsides and exhausted soil in the wake of their westward migration; and lastly, those planters who saw no virtue in the overseer class, to which he belonged. He defended the system of hillside ditching against all charges of "humbug," believing that ridicule had done more to retard agricultural progress than all other things combined.[50] He corresponded frequently with such agricultural leaders as David Dickson, Daniel Lee, Noah B. Cloud, Martin W. Philips, Richard Peters, and Isaac Croom, to whom he referred as old veterans in the cause of his country's salvation. "It has been by corresponding with such men, and reading what they have [written] that I have been aroused and informed," he said. "And the thought of learning more and more on farming and planting and corresponding with such men . . . gives me half the pleasure of life."[51]

Harmon claimed to have studied agriculture "as a physician studies medicine," and to have read all the agricultural works available. He insisted that all agricultural clubs should include in their published reports specific and detailed methods used in producing all premium-winning crops. He encouraged all planters to devote a small portion of their land to an agricultural experiment.[52] "[If] . . . every farmer in Georgia, overseer, or what not, was to devote a portion of his time and lands to experimental purposes, and report the result in an Agricultural journal, such a flood of light as would burst upon the Agricultural world, we have perhaps never dreamed of," he said.[53] On one occasion he took severely to task a planter who reported prematurely on an experiment in planting a single seed of the Japanese pea, which was sent to him from China. "The vine is eaten greedily by stock of every kind, as I tested in a small way last year," the experimenter reported enthusiastically. Harmon observed that with only one seed planted, he must have tested the matter in an exceedingly small way. "I am sorry to see such statements in our agricultural journals," he said. "It injures our cause, and writers should be more careful how they tell their tales."[54]

In 1858 he proposed to be the first of twenty men to give five dollars for a silver pitcher to be awarded the man who made that year the agricultural experiment of greatest value to the South. He proposed a similar award for the best essay on the renovation of exhausted land.[55]

Harmon's ideal of a model planter was one who produced everything he needed for home consumption at the risk of making a small cotton crop. He should also study agriculture as a science and take care of his land as an obligation to posterity. He had no patience with those who saw nothing in farming except large crops of cotton. "The country will be ruined soon enough by politicians and abolitionists, without your aid," he warned.[56] In its agricultural column the *Wiregrass Reporter* described the management of a successful cotton farmer, to whom the editor accorded much praise. Harmon was constrained to offer a rebuke to the editor for calling "a model farmer" this man who had no concern for horizontal culture, rotation of crops, or stock-raising.

"He cultivates thirty acres to the hand, and does it as easily as some planters do twenty," he wrote. "That is what he done [*sic*], and ruined his plantation.... If I had but one prayer to offer for agricultural improvement, it should be, O! Lord deliver my country—my native South—from such 'model farmers.' "[57]

Since the abandonment of the old plantation was a manifestation of a soil-killing culture, Harmon held no brief for emigrant planters. Revealing a sentimental turn of mind, he spoke of such persons leaving their father's grave in the hands of strangers, parting forever with old friends and childhood's happy walks, leaving the old orchard and a cool spring of water to take their march westward to live among strangers, to drink bad water, to clear up new land and then to wear it out in the same manner. "The gullied hills on one side, and the 'fresh lands' on the other will keep the tide of emigration afloat until the South becomes a desert waste," he said. "And then, and not until then, will our children's children commence, alas; commence!! to study agricultural science, and to improve the old, red hills in order to live."[58]

Harmon claimed that the South's system of agriculture was doing more to destroy slavery than all other causes, including the abolitionist movement. "[Our] system of agriculture must be changed, or we are, beyond all question, a ruined people," he said. "Thousands of plantations have been brought from the woods into cultivation within the last fifty years, and worn out, and its owner forced to leave or starve." To him, the criterion of good plantation management was not what it would pay in immediate profits, but what it would pay over a long period of time. "Let us quit moving," he said. "Let the planters of the old cotton-growing districts ... feel themselves at home. Let them fill up the old gullies—improve the old red hills—prune the old orchard—improve the old homestead—enjoy the society of old friends—visit the old moss-covered church, in whose yard slumbers the remains of long departed friends. Then ... will the South begin to grow stronger and her institutions placed upon an immovable basis."[59]

He believed that the poor man most of all could ill afford to wear out his land, that it was most important for him to seek higher productivity on fewer acres. It was poor management to keep up fences around a hundred acres when fifty acres would suffice. He criticized those who claimed they had no time to improve their land, maintaining that they had even less time to wear it out. "But the South will not believe this, until they are crowded together, and compelled to believe it by ocular demonstration," he said.[60]

Harmon's favorite topic was how to keep land from washing away. He looked upon good land as fundamental to progress of any kind. He believed that the Southern Piedmont would never be settled by a permanent population until its lands could be improved and reclaimed. Farmers who only carted manure to their gullied hillsides in a desperate attempt to prolong their productiveness were "eating soup with a fork," he said.[61] A little skill in the use of a simple rafter level would end the greatest spoliation to which the South was heir. Harmon's description of the techniques of horizontal culture did not equal that of Hardwick's for succinctness and clarity, but there was no man who crusaded more eloquently and persistently in behalf of Southern soil than this Georgia overseer.[62]

In January 1856, Harmon severed his connections with Thomas H. Sparks in Vann's Valley, where he had enjoyed an uninterrupted tenure of nearly ten years, an experience unlike that of most overseers. For the next two years he worked for a planter at Utica, Mississippi. Here he suffered attacks of dyspepsia and went for a time to Mont Vale Springs near Knoxville for recuperation. Subsequently he never stayed longer than a year at any place of employment, but apparently this was because planters were bidding desperately for his services. When the *Southern Cultivator* in 1857 began a movement to bring him back to Georgia as a traveling missionary for better soil culture, he informed the editors that his prospects as an overseer were "very flattering" at that particular time. "I have no less than four propositions for next year," he said, "and I must have some encouragement before I can get the consent of my mind to yield up my business and engage in something else."[63]

Beginning in January of the following year, he was engaged by Martin W. Philips, one of the best known agricultural experimenters in Mississippi.[64] The two men had almost identical philosophies on livestock growing. Harmon spoke of their conducting experiments with hogs and of visiting famous plantations in the vicinity. Philips' diary indicates that he had some misgivings concerning Harmon's plan of "pitching the crop." The relationship ended in apparent harmony at the end of the year, however, and Harmon was succeeded by an overseer at the standard salary of four hundred dollars a year.[65]

An interesting prelude to Harmon's employment by the Mississippian was a controversy which the two waged earlier in the columns of the *American Cotton Planter,* in 1854. Harmon took issue with Philips on the latter's theory that overseers should be closely supervised and their prerogatives kept to a minimum. Harmon had very little faith in the ability of the average planter to conserve his own soil, and he maintained that a good overseer should demand complete control of the Negroes and the land. "I would now say . . . to my brother overseers . . . never to agree to oversee for a man who wants you to 'go by directions'; for I assure you that man is mistaken; he only wants a 'driver,' and you will be more troubled by him than forty negroes."[66] But Philips held that it was incredible that a master on hiring an agent could not retain control over his own affairs. Of Harmon, whom he then had never seen, he said with some sarcasm: "I had taken up the idea that friend H[armon] was a man beyond my years, a large, portly, good humored man. I have heard he was a general, and knowing that a Georgia Major was some, I naturally thought a General was more."

In his reply Harmon showed great respect for Philips' contributions to agricultural knowledge, but he did not retreat from his position that an overseer should be given full responsibility for the plantation. ". . . a very common man can manage a plantation better, to be always on it, than a very uncommon man can, to be absent from the plantation more than half the time," he said. He informed Philips, who was also a practicing physician, that he was neither a Georgia major nor a general, "nor am I an old steam doctor, nor do I wish to be either. I am content with the stall I am in."[67]

In this and other controversies which he had from time to time with planters of more or less prominence, Harmon demonstrated his devotion to a cause rather than to an avocation. In reply to a scathing criticism of the class to which he belonged, he advised all overseers to "treat with silent contempt" such criticism of their calling. "Let us avoid everything like strife and let our object be the prosperity of the country," he said. "We profess to be united mutually in the great work of renovating Southern lands, and let us never stop short of its accomplishment."[68]

In Mississippi, Harmon continued his preachments in behalf of horizontal culture. "I have been ridiculed here by some who call themselves p-l-a-n-t-e-r-s," he wrote, "because I contend that a worn out place may be improved at less expense than one can be taken from the woods." He believed that Mississippi planters were inferior to those of Georgia because the former placed too much emphasis on the cotton crop. He found the planters of his Mississippi community to be "enterprising men of capital," but he asserted that not more than a thousand loads of manure were hauled annually into the cotton fields of that county. He demonstrated his method of making compost in the field where it was to be used, and he called it "Georgia manure." He claimed to have located the first hillside ditch in that county. "They have 'crooked' rows . . . but no horizontalizing that deserves the name," he observed. He thought that nothing but a miracle could save that area from dilapidation and ruin. "It is Cotton, Cotton, and Corn, Cotton, Cotton, and Corn, all over the country." He warned prospective Georgia emigrants of the bad water and advised them to think twice before leaving their father's spring of water for Mississippi.[69] He thought that the *Southern Cultivator* had done more to improve Southern agriculture than any journal in the country, and he began a one-man campaign to obtain new subscribers in Mississippi. "I love to read the Cultivator because it is published in the Empire State of the South—the State in which I have lived and labored the last twelve years, and the state which stands higher, in an agricultural point of view, than any other State south of Mason and Dixon's Line . . . ."[70]

From Hinds County, Mississippi, Harmon went to Milliken's Bend on the Mississippi River in Madison's Parish, Louisiana. Here he stayed one year and then went to "Compromise Place" near Tensas Bayou, a few miles farther west. For the first time in his career, he found no need for hillside ditches and contour plowing. "I now see how foolish I was to think that Floyd, Polk, and Cass Counties, Georgia, was the best country in the world," he wrote. "I cannot for the life of me see how it is possible for a country to be much better than this. The soil is at least ten feet deep, [and] the face of the country is as level as old ocean in the profoundest calm . . . ."[71]

He found much of which to disapprove in the fertile lands of the Mississippi Valley, however. He condemned the practice of burning all the accumulated litter in the fields before the spring plowing. He reproached a Louisiana politician for saying that grass was something to be destroyed and not grown, maintaining that a man of good sense and cultivated taste would hardly express in public such unprogressive sentiments. He told of the Mississippi River's going on a rampage and of his attempts to save a thousand-bale crop of cotton. Once he complained of being unable to read at night after the toils of the day "without being bedeviled by forty niggers—here after everything you can mention."[72]

By 1860 this overseer's reputation among readers of agricultural journals was well established throughout the South. "Mr. Harmon is a trump not often turned up from the pack of which he is a member," wrote a Louisiana planter. "Cannot the old state of Georgia send us a few more men of the same stripe to fill the places of incompetent men, who are receiving salaries of which they are unworthy, and which would be better paid to better men?"[73]

In response to a similar expression from Noah B. Cloud, Harmon wrote: "I thank you and others for your kind expressions of approval . . . of my poor services in the great cause of our country's Agricultural salvation."[74]

Although Harmon was the recognized spokesman for the overseer class and often championed their cause against landlords who maligned them,[75] there is little doubt that he cherished a dream of one day becoming a land-owner himself and leading a more serene life. Such a yearning is revealed in the following letter:

> When I was overseeing for Judge Sparks, in Georgia, I thought that the lands in Cedar Valley and Vann's Valley and the lands in Cass County, about Cartersville, was just as fine as I cared to have. Riding along, looking at the plantations and the lands of that country, I have thought many a time, that if I was so fortunate as to own a place there and about thirty hands, I could make it the *model* place of the South. I have located myself upon several plantations there that I could name, and then marked out the "modus operandi" by which I would proceed to make it *the place*. I have horizontalized it, composted it, sub-soiled it, laid off my grass lots and clover lots, arranged my buildings, planted an orchard, set out shade trees, employed a landscape gardener, plotted off my vegetable garden, and then, when all was accomplished entertained my friends, and Oh! how near Paradise I thought I should be, if in that condition . . . .[76]

This letter was written about the same time that Edmund Ruffin, an aging son the the Old Dominion, was witnessing the opening shot against the walls of Fort Sumter. It is to be questioned whether Ruffin's broad acres at Marlbourne made him a more militant Southern Nationalist than the agrarian ideals of this landless overseer. On July 7, 1861, Harmon enlisted as a private in the Ninth Louisiana Infantry. Two weeks later, on the first day of the Battle of Bull Run, he wrote to the *Southern Cultivator* from his station at Richmond: "I received many letters of inquiry before I left home, which I had not the time to answer. Please tell my correspondents that I will be at Manassas Junction in two days, in the midst of the cannon's roar, and perhaps may never reply to their favors. Good-bye old friend Cultivator."[77]

Harmon's military career in the cause of Southern independence lasted for only two years. Its premature end was the result of weakness of the flesh and not of the spirit. From June to October 1862, he lay sick at Staunton and for the rest of that year he was incapacitated for active duty. During the winter of 1863 he was hospitalized at Richmond. In April he was discharged from the service with a surgeon's certificate of disability. "Palpitation of the heart" had rendered him unfit for the heavy marching in the campaigns of the Shenandoah.[78] Back in Georgia, at Marietta, he wrote again to the editor of the *Southern Cultivator,* in April 1863: "Stonewall Jackson kept us moving so, in the Shenandoah Valley last spring, and up to the battle of Fredericksburg, that I had no time to write to you. I have often desired to do so about the beautiful clover fields, grass fields and lawns of Northern Virginia . . . . I have gathered some facts in relation to Virginia farm economy that I hope will be of use to me in after days, provided our cause triumphs . . . ."[79] He asked farmers to grow more peas and beans. "Nothing is so much relished by a soldier in camp as a good kettle of bean soup," he said. Unable to get back to Louisiana, he applied for employment in Georgia. "I have done all I could for near two years with my musket in Jackson's corps; now I want to do what I can with the plow," he said. "I can manage a plantation as well as ever."

He found employment at Sand Town, in Campbell County. There he set about his task of overseeing by day, and by night writing sage letters on the improvement of Southern soil. He never predicted the military defeat of the South but he contemplated her future with forebodings. "Remember we are not to be subjugated by the sword," he wrote. "If we are subjugated at all, it will be done by the plowshare . . . . God alone knows our destiny. Let us plow deep, trust in God and keep our powder dry and the torrent of invasion will be rolled from our own native land and the rainbow of promise once more span our heavens."[80] While Sherman's army was hammering at Georgia's northern passes, Harmon discovered new morals for improved culture and conservation of the soil out of the welter of unhappy events which were slowly but surely changing the direction of Southern agriculture.

At Sand Town Harmon was in the path of Sherman's invading army in 1864, and Hood's evacuation of Atlanta forced him to move again. After the war he worked for several years in Cobb County amid scars of conflict and devastation. Later he moved to Dooly County where he was employed by General Bryan Thomas and where his trail mysteriously ended in 1869. He wrote an occasional article to the *Southern Cultivator,* the only Southern agricultural journal which survived the war, once recommending Chinese labor for agricultural needs. His letters were infrequent, however, and they bore little evidence of enthusiasm for the agrarian life under the new regime. Like Edmund Ruffin of Virginia, the zenith of Harmon's career passed with Appomattox. The elderly Virginian, on hearing of Lee's surrender, had placed a gun to his head and ended his life. The Georgia overseer wrote soon after the war: "I have the inclination, but not the means to till the earth . . . the rose has lost its fragrance—the birds do not sing so sweetly, nor the fields look so gay."[81] Each of these men was an intense Southern nationalist and each had a deep spiritual relationship to the soil.

VIII

Agencies for Promoting Agriculture

WHILE PERMANENT agricultural organizations and a significant agricultural literature had developed in Maryland and the Old Dominion by 1820, such developments did not occur in Georgia until two decades later. Although as early as 1810 the Agricultural Society of Georgia was incorporated at Savannah,[1] its existence was brief, and in 1821 it was noted that Georgia was the only state on the Atlantic seaboard in which there was not a single agricultural society. At this time the Pendleton Agricultural Society of South Carolina had attracted several Georgia members.[2]

Immediately following the 1823 hurricane which wrought great havoc to the growing crops on the coast, the Union Agricultural Society was formed at Darien, including in its membership the more prominent planters of the coastal region.[3] Thomas Spalding's opening address to this group was significantly silent on the subject of wasted soil, but he deplored the "restless desire to roam" which he claimed was affecting adversely existing agricultural practices. The purpose of the society was to bring about the general diffusion of agricultural knowledge among its members and to emphasize subsistence farming while exploring the possibilities of new money crops. Included in the last was the possibility of promoting a revival of such crops as grapes, olives, hemp, and indigo. The society languished for a year or two and then expired for lack of patronage.[4]

Between 1819 and 1830 at least four societies were formed in the Piedmont. A short-lived society existed at Lexington in Oglethorpe County during that period, and there were others at Eatonton, at Milledgeville, and in Wilkes County. These appear to have exercised little influence on their communities and they soon became inactive.[5]

The five-year period following 1835 is significant in Georgia for the inauguration of public measures to encourage agricultural organizations. In 1836, Governor William Schley urged the legislature to provide for a geological survey to explore the state's hidden treasures in mineral and agricultural resources. Ten thousand dollars was appropriated for a geological survey and John Ruggles Cotting was appointed to the office of state geologist.[6] The survey was a resumption of the work of the older Board of Public Works set up in 1825 and abolished the following year, although the earlier emphasis was on transportation alone, rather than on agriculture, mining, and railroads. Sensitive to waning economic conditions, the legislature in 1840 withdrew its funds and the survey was discontinued. Cotting was encouraged to continue his work under private auspices, and he was allowed the use of instruments and equipment possessed by the state.[7]

In addition to providing for the geological survey the legislature in 1837 incorporated a Board of Agriculture and Rural Economy. Composed of forty members, this board was to hold at least one meeting a year at Milledgeville. Like its predecessors, it failed to serve any constructive purpose. In the year following its inauguration, the state treasurer was authorized to pay a bounty of fifty cents a pound for silk cocoons.[8] This effort to revive silk culture on a substantial scale proved abortive, and the subsidy was discontinued after two years. In 1840, however, a little under three thousand pounds of cocoons were reported, Chatham County being one of the larger producing areas.[9]

The most significant movement for agricultural organization coincided with the depression of the early forties. This was a grass-roots movement and stemmed from such problems as soil depletion, low cotton prices, and general financial stringency. Population loss was also a factor. The *Southern Cultivator*, established at Augusta in 1843, was perhaps the most important single agency in promoting the formation of active agricultural societies in many local communities where there was considerable interest in new farm crops.

The new agricultural societies appeared to flourish best in Middle Georgia where out-migration was most common.[10] Here was found a rare combination of gentlemen farmers on very poor soil. One of the more famous societies in Georgia and one which became well known throughout the entire South was the Planters Club of Hancock County. Geography did not endow that community with lasting agricultural assets. It lies in the Lower Piedmont region just above the fall line. In the southern part of the county is much coarse sand and sandy loam, and in the northern end, red clay. In the region just east of Sparta granite underlies the soil at various depths. In some places it sinks deep into the earth; in others there are many acres where there is little or no covering of soil. Today "black jack" oak and pine are not uncommon on the uplands.[11]

Living in this relatively unproductive area and farming its better lands was a group of enterprising planters who made significant contributions to the agriculture of the state. Included among them were William Terrell, the father of agricultural education at the University of Georgia; David Dickson, "the Prince of Southern Farmers"; David William Lewis, the first secretary and one of the founders of the Southern Central Agricultural Society; Richard S. Hardwick, popularizer of hillside ditching and horizontal plowing in Middle Georgia; Dimos Ponce, Carlyle P. B. Martin, Andrew J. Lane, James Thomas, Joel Crawford, Eli H. Baxter, Linton Stephens, Bishop George F. Pierce, Richard Malcolm Johnston, and Algernon S. Brown. Each of these tended to specialize in some novel phase of agricultural improvement. Thomas, Dixon, and Brown became specialists in labor management. The last owned a tract of land outside Sparta which he converted into a colony for superannuated slaves.[12]

Fully two-thirds of all the local societies in Georgia were found in the old plantation counties of the Middle Georgia Cotton Belt, more than 90 per cent of which were organized between 1840 and 1845. With a few exceptions the remaining organizations were in southwestern Georgia and in Cherokee Georgia.[13] The societies in the western part of the state generally were not formed until the 1850's. In contrast to the pattern in the older counties, the western societies were concentrated in the more fertile areas, and they emphasized soil conservation rather than new or non-cotton enterprises.[14]

Those who took the lead in the agricultural societies were invariably men of wealth and social position. Ex-Congressman Joel Crawford wrote the constitution of the Hancock Planters Club, which provided a model for the organization of other societies. Later Crawford moved to a plantation in Early County and organized the first local club in southwestern Georgia.[15] John M. Berrien, former United States attorney general, was the first president of a society organized in Chatham County in 1845. Ex-Governor Charles J. MacDonald was the first president of the Agricultural Association of Cobb County. Bishop Stephen Elliot was president of the Central Horticultural Society organized at Macon in 1849.[16] Other founders and leaders include Dr. Philip Mims in Murray County, S. L. Pope in Floyd, John W. Moody in Oglethorpe, Thomas M. Berrien in Burke, James Camak in Clarke, and James R. Wyly in Habersham.[17]

The Hancock Planters Club, which provided a general pattern for subsequent organizations in Middle Georgia, was motivated by a strong civic aim as well as the improvement of individual fortunes. The club was organized in 1837 with eighteen members and elected Dr. William Terrell its first president.[18] Among its early members were ten prominent planters, four county officials, two prominent school teachers, and a judge of the Superior Court. On its subsequent rolls were names of the leading men in the county, among whom were Judge James Thomas, Methodist Bishop George Foster Pierce, Dr. Edmund Monroe Pendleton, Linton Stephens, David W. Lewis, and Richard Malcolm Johnston. Nearly all of its members were either planters or professional men with considerable planting interests.[19] John T. Martin, a blacksmith, was the non-planter in the group, but he possessed property valued at more than thirty thousand dollars.[20]

Men of modest economic circumstances were by no means barred from membership, but active participation in the club's proceedings seems to have been confined to men of the planter class. It is significant to note that David Dickson, who became the most successful and the best known planter in the county, was never a member of the club, although his brother, Thomas J. Dickson, played a leading role. David Dickson suffered some social ostracism because of his Negro mistress who was the mother of three mulatto children.[21] Occasionally he betrayed a note of ridicule in reference to the local club. "I do not plant for premiums, or silver cups," he once said, "but for good dividends and plenty of corn to feed my friends' horses when they honor me with a visit."[22]

From the date of its origin to 1841 the society was relatively inactive, and no new members were added to its rolls during this period. In 1841, however, apparently as a result of the depression, fourteen new members appeared. By 1847 it had recorded a total of eighty-one members, many of whom were men of only moderate circumstances. (Fifteen of this total had either died or moved out of the county). After 1841 the society carried on several enterprises. It conducted a campaign to disseminate agricultural literature, held exhibitions of livestock and field crops, and awarded premiums for outstanding achievements. Reports were read on experimental methods, and a county-wide agricultural census was conducted. In 1844-45, the club attempted to inaugurate a south-wide reduction of cotton acreage in order to raise the price of the staple and to encourage planters to devote more time to other crops.[23]

The biennial meetings of the club early gave away to monthly meetings, at which agricultural topics were discussed. These topics varied with the season and with the passing years. In the spring of 1842, for example, a typical query was whether "land that will now produce two Blls. of corn per acre . . . [can] be made to produce within ten years ten bbl. corn per acre [sic] without the aid of manure, other than what can be raised on the land [with] a crop of small grain or corn to be taken from the land every year." Late in the summer of the same year the club discussed the question, "whether it is beneficial or prejudicial to the growing crop to plow the land in very dry weather." The subject announced for one meeting was "the best method of gathering, cleaning and preparing cotton for market with a view to commanding the best prices."[24] Emphasis was placed on subsistence crops in 1845, and the merits of both intensive and extensive farming were discussed. The favorite method of presenting a question was in the form of a debate, a method generally used elsewhere. Indeed the agricultural society at Sandersville began as a local debating club.[25]

By 1860 the annual fair of the Hancock Planters Club had become famous throughout a wide area. Its premium lists were among the largest in the state. For livestock shows it had constructed a large indoor amphitheater with exhibition rooms underneath the seats. The large crowds in attendance could now hear agricultural addresses and witness other proceedings in comfort. The fair was the greatest festive occasion held in the community, unequaled by Thanksgiving Day and the Fourth of July. Great pageantry sometimes marked the occasion.[26] At the 1845 fair, for example, the closing day witnessed a great parade of club members and citizens, escorted by the county militia. The procession formed at the courthouse and marched across the town to the Female Academy where the final address and the awarding of premiums took place.[27]

The premiums awarded by the agricultural societies covered every variety of activity on the plantation. The Hancock club's prizes for 1845 ranged from fifty cents for "the best piece of cotton diaper" awarded to Mrs. E. L. B. Hall of Putnam County, to a silver cup awarded to Richard Hardwick for the best acre of upland corn. In 1843 the jack, the jenny, and the mule first appeared on the premium lists.[28] These lists varied widely from one community to another throughout the state, being largely dictated by the agricultural interests prevailing in the locality. The Agricultural Club of Lower Georgia (Savannah) emphasized household and mechanic arts, awarding premiums for homemade harness, shoes, castings, cannon balls, agricultural tools, sugar, soap, bread, wine, butter, vinegar, knitting, and only a few prizes for agricultural crops.

The Wilkes County Society, in the upper Piedmont, confined its premiums to horses, cattle, and swine, while the Monroe County Society offered its best premiums each for upland and bottom corn. To emphasize diversification and domestic manufacturing the latter society included in its premium list silk cocoons, silk products, and Negro clothing, such as blankets, bonnets, and hats. The Cass County Agricultural Association in Cherokee Georgia emphasized horses, cattle, and sheep.[29] The premium list of the Muscogee-Russell Agricultural Society (Columbus) in 1851 included essays on agricultural subjects, and prizes were given for the best poultry, farm implements, butter, cheese, syrup, sugar, fruits, flowers, vegetables, wares, wine, and articles for home use.[30] The Meriwether club in western Georgia offered a premium for the best-arranged and best-conducted plantation. While it included a limited list of diversified products, it put the main emphasis on cotton and corn, a policy directly opposite to the one prevailing in the older soil-depleted counties.[31]

The Etowah Agricultural and Mechanical Association (Rome) in 1852 had more than forty items on its mechanical list in addition to numerous items in horticulture, floriculture, domestic manufacturing, cooking, preserving, field crops, and livestock. The fair was held at Waleska, the plantation of Judge John P. Eve. The railroad from Rome to Kingston transported all exhibits free. Liquor, dogs, and "unruly animals" were barred from the fairgrounds. Officers of the association wore white rosettes to designate their rank, while members wore blue ribbons, and their servants wore yellow hatbands.[32]

In order to coordinate agricultural activities in the various county organizations a state agricultural society was organized at Milledgeville in 1845, known as the Agricultural Association of Georgia. It named Governor George W. Crawford as president. Members of all local societies were entitled to membership in the state society, whose funds were to be raised by voluntary contributions. At a meeting in November following its organization, thirty-five delegates were present from eleven counties, Hancock having the largest representation.[33] They accepted an invitation to meet at Sparta in 1846 as guests of the Hancock club at its annual fair. But this state organization did not flourish. In 1847 its state fair, at Milledgeville, proved extremely disappointing, one reason being, perhaps, the general inaccessibility of the Georgia capital. The only livestock exhibited was one sow and a litter of pigs.[34]

In the meantime, John W. Graves, seeking patronage for his hotel at Stone Mountain, then at the northern terminus of the Georgia Railroad, invited some prominent Georgians to assemble at that place to discuss worthy non-political issues. Mark A. Cooper suggested an agricultural meeting and he furnished Graves with names of men who might be interested. Several of those named gathered in 1846, and the Southern Central Agricultural Society subsequently was organized. Originally composed of sixty-one members and with a treasury of sixty-one dollars, it held its first fair in 1846. A jack and a jenny grown on the property of Graves was the extent of the first exhibit.[35] After this meager beginning, however, the project prospered. The exhibits the second year nearly filled the tenpin alley across the road from Graves' hotel, and a crowd estimated at three thousand persons taxed the accommodations of the little village for four days. In 1851 the receipts from the fair amounted to $2,000, and there was a two-thousand-dollar subsidy from the city of Macon, to which place the annual meeting had been removed. In 1852 the society was able to employ Joseph Jones as a full-time secretary.[36]

At the time the Southern Central Society began there was not another state or regional society in any of the states adjoining Georgia. It was thought that the agricultural interests of these states could be joined in a strong regional organization, as the name of the society suggests. For seven or eight years this idea promised success and the society enjoyed a nominal membership from other states, but it never became a truly regional organization except in name. In 1855 state societies were formed in Alabama and in South Carolina and these were soon followed by one in Tennessee, all of which appear to have been influenced by the original Georgia movement.[37]

Before the relatively slight reductions made in its membership by newly-formed neighboring associations, the Southern Central Society had approximately a thousand members. At its most prosperous period its revenues from all sources amounted to ten thousand dollars annually. Among other things, it was able to publish, in 1852, a compilation of its *Transactions*, the first of its kind ever published in the South. By 1859 the annual revenue was reduced to two thousand dollars, and the society was periodically applying to the legislature for financial aid. David W. Lewis, who served for nearly ten years as secretary of the society, wrote despondently of the impending failure of the organization. "The cry of 'Book Farming' and all other terms of ridicule [has been] heaped upon us," he said, "but my confidence in . . . the final [defeat of] . . . the opposition is full, perfect and invincible." He regretted the inability of the society to publish another volume of its *Transactions*, to sponsor an agricultural library and museum, to collect implements and objects of natural history, to conduct an experimental farm, and to inaugurate a program of agricultural education. "But the blindness of economy is blind to everything," he said.[38]

Other factors than the blindness of economy may be attributed to the waning influence of the Georgia society and the reluctance of the state legislature to come to its aid. Important among these were several splinter agricultural organizations which developed in the 1850's, and, more significantly, the schism which developed within the ranks of the parent society itself. Those interested more or less exclusively in horticulture were the first to form a separate organization, arguing that the annual fair of the Southern Central Society came too late in the season for a creditable exhibition of fruits, vegetables, and flowers. Composed principally of men living at Macon and Milledgeville, the Central Horticultural Association was organized in 1849 at the former place. Bishop Stephen Elliott, Judge Iverson L. Harris, Simri Rose, Robert Nelson, and James A. Nisbett were among those instrumental in its formation. A number of ministers were members. An exhibition held in July consisted of vegetables, flowers, peaches, grapes, and other summer fruits. The society was incorporated in 1850 and while its name suggested regional ambitions, it was never more than a local organization.[39]

In 1853 Bishop Elliott, now living at Macon, William N. White of Athens, William H. Thurmond of Atlanta, and Jarvis Van Buren of Clarksville formed at Athens the Horticultural Society of Georgia as an auxiliary of the Southern Central Agricultural Society. In the meantime a horticultural society was formed at Augusta which absorbed the interests of a great majority of fruit growers in that vicinity.[40] In 1856, however, the Athens society, which had reorganized as the Pomological Society of Georgia, had become an organization worthy of its name. Louis E. Berckmans, the famous Belgian horticulturist, then living at Augusta, was its president. Richard Peters of Atlanta was its vice president, and William N. White of Athens, its secretary. The display of fruit at Athens in 1858 was a notable success, including 568 lots and comprising numerous varieties in which peaches led with nearly a hundred. The organization sent delegates to represent Georgia at the meeting of the American Pomological Society in New York.[41]

Another short-lived organization which branched from the Southern Central Society was the Mechanics Institute of Georgia, which held a fair at Macon in 1852 in conjunction with the parent society. Its exhibits included thirty-two agricultural implements, twenty-one pieces of machinery, twenty-one steel and iron implements, twenty-five wooden pieces, seventeen articles made of leather, twenty-seven chemical articles such as oils, cements, and paints; and fifteen exhibits of paper, book bindings, and architectural designs.[42]

In addition to the formation of the splinter organizations, there appeared a bitter schism within the Southern Central Society itself. It was led by a faction which sought to restore the emphasis on cotton, since prosperity had returned to that staple during the 1850's. The majority of the membership of the society seemed to favor a program designed for the small planter with emphasis upon diversification and economic independence within the farm establishment.[43] In this group were such men as Richard Peters, David W. Lewis, Charles W. Howard, Thomas Stocks, Mark A. Cooper, William Terrell, Charles A. Peabody, Dennis Redmond, and William Schley.[44]

The leaders of the cotton-emphasis group were Robert Toombs, Howell Cobb, Nathan Bass, James V. Jones, Thomas Butler King, John S. Thomas, Daniel Lee, Joseph Jones, and others. This group appeared to be dominated largely by men of political influence in the state. In 1859 they formed a Georgia Cotton Planters Convention somewhat after the fashion of the society from which they were seceding. They planned a fair of three weeks' duration to be held at Macon in the heart of the cotton region, and in competition with the Southern Central Society.

Howell Cobb was elected president of the new organization and a program was immediately launched for the purpose of securing direct trade with Europe. To negotiate the details, they allocated one thousand dollars in 1861 to send Cobb abroad. Concerning the direction which the new organization was taking, a correspondent to the *Southern Cultivator* asked: "Is it an agricultural movement or a commercial one, or a compound of both? If the latter, I fear for its success, having already witnessed the rise, progress and downfall of more than a few 'Commercial Conventions.'" This correspondent saw no reason for a new organization competing for patronage with the agricultural society.[45] About the same time Charles W. Howard wrote in the *South Countryman* that "in certain important quarters, there is, at present, a want of cordiality toward this [Southern Central] Society. The great bulk of the cotton interest is measurably withdrawn from it . . . ." He deplored the party strife in the legislature which prevented a sober consideration of the needs of agriculture, and he suggested an "active and eloquent agent" to lobby in behalf of the organization, which at that time was petitioning for a small appropriation for carrying on minimum activities. "Mere politicians do not like 'the two blades of grass,'" he said. "They smother other plants which they like better." He asked for harmony between the two organizations in order that state aid might be secured for both. Daniel Lee, editor of the *Southern Cultivator,* was accused of supporting the interests of the schismatic group, and afterwards, in 1859, Howard succeeded him as editor. In general, other agricultural periodicals did not participate in the controversy.[46]

The Georgia agricultural press was among the more progressive and respected in the entire South. Between 1840 and 1860 five agricultural journals were established in Georgia, four of which had notable careers. In March 1843, the *Southern Cultivator* was founded at Augusta by James W. Jones who edited it for the first two years as a bi-monthly quarto of eight pages. Beginning with the third volume, James Camak of Athens assumed editorial charge, and the periodical was converted into a monthly of sixteen pages. James W. and W. W. Jones continued to be the publishers. On the death of Camak, in 1847, the editorship was tendered to Daniel Lee, who remained in the position for more than a decade. Lee formerly had been connected with the *Genesee Farmer* of New York, and he came to Georgia with a wide reputation as an agricultural writer. In January 1851, the *Soil of the South* was launched at Columbus under the joint editorship of Charles Alfred Peabody and James M. Chambers, with a subscription list of two thousand for the first number.[47] Since Columbus was located on the fall line between Georgia and Alabama, the periodical drew much patronage from the latter state. In January 1853, Dr. Noah B. Cloud of Montgomery began the *American Cotton Planter,* which drained much of the patronage from the Columbus journal. Beginning in January 1856, the two publications were merged under the joint editorship of Cloud and Peabody, the latter retaining the title of horticultural editor. At this time the *Soil of the South* was in such decline that the final volume contained largely exchange contributions and re-written articles from other journals.[48]

The *South Countryman* was Cherokee Georgia's contribution to the field of agricultural journals. Published at Marietta, it began in January 1859, under the editorship of Charles W. Howard of Kingston. In the introductory number, the editor announced that the *Countryman* would give more attention to the subject of cultivated grasses than any other journal in the state. Like its predecessors it began with high expectations of a long subscription list. "There are nearly, if not quite 100,000 voters in Georgia alone," said Editor Howard. "[But] there is but one agricultural paper in the state."[49] After a year of crusading in behalf of diversified economy, based upon livestock and grasses, Howard transferred his zeal and enthusiasm to the pages of the *Southern Cultivator,* and the *South Countryman* went the way of the Columbus journal. When he replaced Daniel Lee as editor of the *Southern Cultivator*, the latter became agricultural editor of the *Southern Field and Fireside,* a weekly journal also published at Augusta. Established in May 1859, it devoted much space to an agricultural department, first under the editorship of Lee, and later under Dr. William N. White. It was not wholly devoted to agriculture, however, for it shared its columns with a literary clientele. Of these four journals the *Southern Cultivator* was the only one which survived the Civil War. It was discontinued in 1935.[50]

Throughout the 1850's there was a growing emphasis on agricultural topics in the daily and weekly press. Agricultural and horticultural departments appeared in various papers, notably the *Columbus Enquirer* and the *Georgia Citizen* (Macon). Most of these departments were well edited. Robert Nelson, for example, who edited the department in the *Georgia Citizen*, succeeded Peabody as horticultural editor of *The American Cotton Planter and Soil of the South* in 1858. "The star of the farmer is on the rise," said the *Southern Literary Companion* at Newnan in announcing its Farm and Garden Department in 1860, edited by I. N. Davis. "To be a distinguished man now-a-days there is no safer or more substantial way than to be an eminent agriculturist, successful horticulturist or the like . . . . There is no way for a man to be 'looked up to' for the next half century, like being an enterprising and successful farmer, and there is certainly no way to pass life so pleasantly."[51] By 1860 agriculture had found a place in the columns of nearly all the weekly papers of the state.

The *Southern Cultivator* was recognized as among the outstanding agricultural journals in the country. Its patronage area extended from the Carolinas to Texas. Beginning with less than two thousand subscribers, it had a circulation of ten thousand ten years later, and upon the outbreak of the Civil War it was flourishing.[52] During the greater part of this period its editor, Daniel Lee, was the only high-ranking agricultural editor in the South whose entire experience and background were acquired in a region outside the section. Arriving in Georgia in 1847 from New York where he had been an agricultural teacher and an editor, he quickly conformed to his adopted section's attitudes on slavery and other leading social and political dogmas. He generally kept his journal free of political discussions, but he permitted open discussions of slavery, which institution he came to favor over free labor. He supported the large planters in their agitation for the re-opening of the slave trade, on the ground that the South needed more labor for reclaiming its worn-out lands. Policies which he advocated included state aid for agriculture, public-supported agricultural education, and the reforestation of denuded plantations.[53]

Trained and experienced in a school of diversified economy in the North, Lee came to Georgia at a time when that philosophy was acquiring considerable respectability in the South, even among the most conservative cotton planters. He pursued this policy for a time, but like a wise politician, he charted his course from the signs he read in the market place. Improved cotton prices during the 1850's appear to have encouraged him to adopt the views of the large cotton-planting interests. He lost some influence and a few subscribers by his defection. "When you was *first* editor, you lectured your readers on domestic economy," wrote a correspondent. "Then you was on the right track .... But now I see you have diverged or run into the popular breeze, or cotton mania, which all old farmers know has ruined Middle Georgia." The writer maintained that the "Empire State of the South" based on cotton was "all a humbug" and advised Lee to abandon his course and tell his subscribers how to make something to eat and how to practice a more self-sufficing economy. The letter concluded with the admission, "I think more of you than any Northern man I ever saw." The *Southern Cultivator* remained always the organ of the Cotton Belt.[54]

The career of Charles Wallace Howard, founder of the *South Countryman* and Lee's successor in 1859, was in many respects in direct contrast to that of Lee. A native Georgian, educated at the University of Georgia and at the Princeton Theological Seminary, he was one of the founders of Oglethorpe University at Milledgeville and a member of its faculty until 1839. He possessed means for travel and study, and he had a firsthand knowledge of Southern agriculture. He wanted nothing less than a complete agricultural revolution in Georgia. In a region steeped with all the traditions of cotton and corn culture, he advocated grasses, livestock, crop rotation, and a diversified economy. The return of prosperity to cotton farmers in the 1850's served only to intensify his crusade rather than to weaken it, and he was never accused of deserting to the enemy. He was a pioneer in agricultural leadership, rather than a commentator.[55]

In direct contrast to Howard were a number of men claiming to be agricultural specialists and reformers but who turned out to be commercial shysters. Among these was "Professor" Charles Baer who came to Georgia from Maryland shortly after 1840. Claiming to have a process of making fertilizer from common materials, he offered to apply his product to land at the attractive rate of ten cents an acre. At the state fair in Augusta in 1853 he lectured on agricultural chemistry, announcing later that he would be available as a lecturer to local agricultural societies on renovating worn-out land, provided they offered him sufficient inducement.[56]

In a lecture before the Hancock Planters Club in 1860 Baer stated that "smut in wheat was not in any sense a disease," but was caused by the action of winds and moisture. The validity of this statement was challenged by some members of his audience who asked him to explain, in the context of his theory, why bluestone relieved the condition of smut. On another point his audience pressed him for an explanation of his statement that "roots of cereals did not penetrate the sub-soil of red lands and luxuriate in the salts beneath." Baer's explanation was the presence of "a peculiar acid" in the unplowed subsoil which ate off the roots of these plants. "We have some doubts of his being or having been, a professor in a college, while he talks so much about agriculture," wrote a member of the Hancock intelligentsia. "Will our Baltimore agricultural exchanges please set this matter right before the public? Mr. Baer's theories strike us as being equally crude and unreliable." The results of the investigation of Baer's background are unknown, but he made a sudden exit from the ranks of Georgia's agriculturists, after a career of several years.[57]

The farmer's gullibility and the frequency with which he was victimized by such unscrupulous peddlers helped to focus attention on the need of public support for agricultural education. The movement to reorganize educational policies in Georgia in the 1850's was to a great extent an outgrowth of the agricultural reform movement in that period. The democratization of secondary education through the inauguration of a public school system and a change of emphasis in higher education both were reflected in the proceedings of the Southern Central Agricultural Society. At a meeting of this society at Macon in 1850, the Reverend Thomas F. Scott of Columbus delivered an address in which he stated that there were forty thousand children in the state whose parents were unable to pay for their education. He gave an eloquent argument favoring a state-supported system of common schools. "This blessed consummation," he said, "would do more than anything . . . to raise up and fix upon our soil a permanent, home population."[58] Apparently responding to Scott's address, the executive committee of the society called for a convention to meet at Marietta in July of the following year to inaugurate the new educational movement.[59]

Agricultural leaders also were concerned for higher education, the curriculum for which failed to serve the interests of agriculture and the agrarian life. Instead of creating better farmers, the colleges were accused of educating the sons of planters away from the soil and sending them into the overcrowded professions. Charles W. Howard thought that the state should endow an agricultural school designed to train young men to become common school teachers and effective farm overseers.[60] Much criticism was directed at "literary institutions" which did not emphasize the "practical sciences." Carlyle P. B. Martin believed that youth should be taught by "ocular demonstration," and Charles A. Peabody maintained that the study of nature was worth more than "all the liturgy of a trained pedagogue."[61] Jethro Jones thought the literary curriculum of colleges was responsible for the Southern attitude that all manual labor was menial and revolting. He called such institutions "hot beds of dress and droneism" and their products he termed "a class of professional drones, yawning around our courts of justice, and the manufacturer of pills and bills at every shop and cross-road in the land . . . ." With a sly reference to the Georgia Military Institute recently started at Marietta, he called for public-supported schools "for educating our youth in the use of the plow, as well as the sword."[62]

One planter thought that Latin and Greek were useless appendages to an overweighted curriculum and suggested that German be substituted in the hope that a student might "learn something agriculturally" from the modern Europeans. Joseph Henry Lumpkin reminded the critic of classical learning that "Homer and Hesiod . . . sung the praises of the plow, [that] Virgil . . . collected the best observations and choicest maxims of antiquity on rural occupations; that Cato, Varro and Pliny and their contemporaries, speak familiarly of . . . preparing the ground, sowing seed, cultivating the crop, reaping, harvesting, threshing and . . . feeding, clothing, and lodging agricultural slaves . . . and selling all kinds of cattle and poultry, [and] the employment of overseers . . . ."[63]

The movement for agricultural education was originally designed to modify the type of instruction given in the academy schools. In 1845 Alphonso Rogers, a student at the Warrenton Academy, delivered a school oration in which he sought to call attention to this problem. "Why have many splendid habitations been deserted, and why do the owl and bat revel in these halls once vocal with the sound of mirth and prosperity?" he asked. "No more are heard the woodman's axe and plowman's song among the hills, once presenting the appearance of abundant vegetation, but now of red clay." The remedies which he listed included the inauguration of agricultural courses in schools and academies.[64]

In 1845 Carlyle P. B. Martin announced the opening of his "Family Boarding School" at Madison and informed the public that in connection with the usual studies, "lectures on agricultural chemistry will be delivered before the students, and the principles of Chemistry as connected with the noble and important subject of agriculture illustrated by experiment and analysis." Later Martin operated a similar school in Henry County. Profiting from the experiences of an earlier day, he steered clear of the older manual labor idea in the organization of his curriculum. "No manual labor is performed by the pupils," he said, "but the whole spirit of the Institution is eminently practical, particularly as it relates to subjects of interest in plantation life."[65]

In less than one year the school was filled to capacity, and forty pupils were refused admission in 1857. When the crowded condition became generally known, friends of the movement solicited donations throughout the South to enable the institution to expand, claiming that it was the only school of its kind in the South. Martin negotiated for the purchase of the 600-acre farm and buildings at Montpelier Springs, in Monroe County, formerly occupied by the Female Episcopal Institute. In soliciting support for agricultural education, I. N. Loomis of Macon said:

> The day is not distant, when our sons will be in the far west, and the homes of our fathers in the possession of the alien and the stranger! Melancholy thought! Their bones may be turned up by the plowshare of the foreigner, and their ashes trampled beneath his unhallowed feet .... Again we ask, will no other heart beat responsive to Mr. Martin's aspirations for the promotion of Agricultural Education? We believe there are many who will cordially second his noble endeavor, by aiding him to the means of procuring Mount Pelier, where his great plan may be fully embodied on the grandest scale.[66]

Such was the publicity accorded Martin's agricultural academy. Farm machinery was donated to the school by various firms, one as far away as Baltimore. David Dickson sent a sack of select cottonseed. Richard Peters promised two Devon cows. Plows were sent by a manufacturer at Thomaston. In 1860 the school was in a flourishing condition, but it soon was to succumb to the Civil War.[67]

The educational movement begun at Macon in 1850 was climaxed eight years later when the legislature provided for the setting aside of $100,000 annually from the rental of the Western and Atlantic Railroad to be used in financing a public school system in Georgia.[68] William Terrell's gift of two thousand dollars to the University of Georgia was said to have established the first liberally endowed chair of agriculture in the United States.[69] His gift was inspired by the prevailing sentiment among agricultural reformers for the introduction of agriculture into the curriculum of colleges. The legislature was urged to appropriate a sum to match the Terrell gift, but that body remained indifferent to such appeals. The *Macon Journal and Messenger* said in 1858 that

> Georgia has fostered and patronized Medical Colleges, and the last legislature authorized the purchase of a Military Institute, to educate her sons in the arts of war—but what has she ever done to give an intelligent direction to the plow share and the pruning hook—to the implements of peaceful and productive industry?
>
> Under this fatal policy, Southern soil is wasting away and our white . . . productive population constantly diminishing. Our fertile lands are falling into the hands of a few large slave-holders, whilst the small planter and the mechanic seek new fields of labor and enterprise . . . . Let agricultural schools be established and fostered—let every species of honest labor be dignified and ennobled.[70]

Thus it was that the agricultural reformers in the decade before the Civil War laid the foundation of a public attitude favoring a new type of educational curriculum. Its ultimate realization in Georgia came in the following century.

IX

The Quest for Grasses and Improved Livestock

LIKE OTHER states of the Lower South, Georgia had an abundant supply of range land for livestock but was notably deficient in natural growth for fall and winter grazing. The deficiency was particularly true of the Piedmont Cotton Belt where unsuccessful efforts to supply winter grazing had been undertaken soon after the region was opened for settlement. Analyzing the failures in 1830, William Terrell stated that they resulted largely from the cotton grower's natural aversion to grass, his unwillingness to prepare the ground properly, and his tendency to assign worn-out cotton fields to pasture lands.[1]

Terrell was at that time experimenting with Virginia lyme, a winter grass which he found would not re-seed. Cattle grazed it so closely that, when the earth was wet and loose from winter rains, they pulled it up by the roots. A few enterprising individuals continued to experiment, among whom was James Thomas of Sparta. He claimed to have tried all grasses that came to his notice, including samples from Kentucky, Texas, Ohio, and Pennsylvania.[2] N. B. Moore of Augusta succeeded in growing successfully such crops as Timothy, red clover, Lucerne, Scotch, and Italian rye grasses. William J. Eve, Jarvis Van Buren, Edwin Williams, and George H. Waring, all of whom lived in northeastern Georgia, demonstrated that grass could be grown with some success in the mountain region above the Cotton Belt.[3]

Knowledge concerning these experiments was difficult to disseminate because of the confusing nomenclature of grasses. For example, what later came to be generally known as Bermuda grass was identified variously as doub grass, wiregrass, Joe Jointer, and Tarleton grass. The last name originated from the belief that it was first brought to the South during the Revolution in the saddle padding of General Tarleton's troopers. What Southerners called Timothy grass New Englanders knew as Herd's grass, and what Georgians identified as Herd's grass New Englanders recognized as red top clover.[4] A favorite grass in Clarke County in 1860 was known locally by five different names, these being oat, Standford, Utah, meadow, and Smyth grass. Johnson grass was also known as Means and Guinea grass. Virginia lyme grass was known in many sections of Georgia as wild rye. Because William Terrell of Sparta brought it prominently into notice around 1830 it became known as Terrell grass throughout Middle Georgia. It was also known both as deer park and mesquite grass, depending upon the region in which it was grown. As early as 1832 alfalfa was known as Lucerne, Chilean clover, and French clover. As late as 1860 many planters thought the three names represented three different plants.[5]

Confusion in nomenclature combined with general ignorance of botanical science was a source of many frauds in which farmers were victims. Typical was a claim by B. V. Iverson of Columbus that he had acclimated a winter grass for Georgia conditions. Because it appeared to answer all the requirements of a grass which Southerners had long desired, Iverson called it "the Rescue grass of the South." Taking advantage of the widespread wish to convert red and eroded fields into carpets of green pasturage, Iverson sold the seed for five dollars a peck, with an obligation on the part of the purchaser to grow none for sale.[6]

Iverson's "Rescue Grass" had the run of the Lower South for about three years, until the accumulated record of its failures over a wide range of territory brought it to grief. The fact came out that the seed of the "Texas oat grass" had been planted by Iverson in his garden at Columbus in 1851 and, although an annual, it survived during the unusually mild winter of 1852 when second-growth cotton was reported at many places. Iverson then bought a large supply of the seed and made preparations to supply a growing list of customers. The extent of speculation in this grass was enormous. One seed dealer bought over two hundred bushels at ten dollars a bushel and retailed the seed in pint packages at the rate of $160 a bushel.[7] It was finally discovered that this far-famed grass was actually the chess, *Bromus secalinus,* and for more than a century afterwards it was known as cheat grass.[8] "Were I gifted with wit," observed one farmer, "I would puff the Crab Grass, and advertize the seed at a dollar a teaspoon full."[9]

Only a few individuals devoted any sustained effort to improved grasses before 1850. Daniel Lee, who had long experience among livestock farmers of the Genesee Valley in New York, studied carefully the possibilities of such an enterprise in Georgia. At Augusta in 1848 he noted that the indigenous grasses of that section did not begin to grow until after Timothy, wheat, rye, and barley were ripe, and he believed this to be an advantage. After giving a lecture to local planters on the subject of growing grasses, he found his audience amused at his seriousness. "How to destroy grass, not to increase it, was the information they wanted," he noted afterwards.[10] Cotton and corn, being intertilled crops to be kept free of grass and weeds, caused their cultivators to hold grass in utter contempt. After despairing of his attempt to get Southerners to grow grasses, Lee tried to persuade Northern farmers to come South and engage in the livestock industry. "For a crop of hay and land is plowed in June, harrowed, and *not* seeded," he informed the readers of the *Genesee Farmer* of New York. "The grass is mown about the first of August, and yields from one to three tons of good hay, according to fertility.... A crop of peas is made after early corn.... A Northern dairyman could make a fortune here by making butter and cheese."[11]

The importance of grass, both as an aid to improved livestock and as an element in a scheme of crop rotation continued to be emphasized by leading planters after 1850. Since there was no native grass which met all the requirements, planters were urged to try Bermuda in the Georgia Piedmont. This grass had long been used for holding the soil on the levees on the lower Mississippi, and there Thomas Affleck early became familiar with its possibilities. Thomas Spalding also found it a valuable grass on Sapelo Island, and both men recommended it for the Georgia Cotton Belt where it became well known about 1840[12] At this time it was believed to offer a solution to the problem of soil erosion.

Bermuda grass had an advantage over most other grasses in that it could survive heat and drought. Other grasses would invariably succumb to the heat of the late summer and fail to re-seed. However, Bermuda stems became dry in the late summer and were uninviting to livestock from about the first of August to April. Imported winter grasses had such a shallow root system that they could not be grazed without destruction of the plant. On the other hand, the extreme tenacity of Bermuda and its resistance to eradication made it undesirable to use in a system of crop rotation.[13]

No sooner had Bermuda grass become established in the Piedmont than there arose a vigorous cry for its immediate and complete extermination. The *Macon Telegraph* stated that if Spalding and Affleck could furnish an answer "whereby Bermuda grass may be effectually rooted out . . . they will confer a favor on the planters of the South that will not soon be forgotten." John W. Pitts, of Newborn, after several years' experience, said that it was destined to ruin the country unless the law interposed a barrier to its dissemination. "Once rooted . . . on any portion of a plantation, it is as good to take that plantation as a sevenpence is for a half pint," he said. He cited many plantations in Greene, Hancock, and Putnam counties which had been rendered worthless by this grass.[14]

John Cunningham was not sure whether it was a blessing or a curse. "As to killing it by any kind of plowing, or planting hoed crops among it, this is utterly impossible . . . . I would as soon try to drown a fish by throwing him in the water, as to kill Bermuda Grass by never [sic] so much working in the sun." A Jefferson County farmer said that the "fallacious flatteries . . . and false praise that occasionally adorns its ignoble character" came from those who had the grass on their plantations and did not want their property depreciated. "[They] desire to sell out and quit it, since there is no way to get rid of it on fair terms," he said. A Newton County farmer proclaimed that a farm stocked with Bermuda was the worst curse a father could pronounce upon his son. John Farrar of Atlanta offered $500 to be entirely rid of this pest on a lot of fifty acres. A South Carolinian in 1848, traveling along the old state route from Washington in Wilkes County through Greensboro and Eatonton to Milledgeville, observed a number of farms with fine dwellings and improvements entirely deserted by their owners. "On inquiring the cause," he stated, "the evidence was abundant that Bermuda had actually rooted them out . . . . So the doors of the mansion were thrown open, the gates around the fields let down, and the owner set out for the far West."[15]

As time passed Bermuda came to be tolerated by intelligent planters who developed methods of controlling it, but the average farmer abhorred its presence anywhere on the plantation. John

W. Rheney of Burke County learned to plow it in winter thus to expose the roots to frost and then to work it with a harrow. John Cunningham finally resorted to planting Bermuda fields with over-shadowing crops and then to plowing up the grass during the dry seasons of August and September. He was thus able to use it in a rotation system with limited success. Later, he referred to it as "that blessed pest." Edmund M. Pendleton called it the "grass of the South," while Richard Peters considered it to be as reliable as any other grass. Francis Ticknor recommended it for town commons, cemeteries, old fields, lawns, and "Grove-surrounded nooks" where it might be at peace. "The profane wretch who pours contempt upon it, should be compelled to eat May-Pops and drink Florida Coffee the rest of his unnatural life," he said.[16]

In Hancock County, Bermuda acquired its more steadfast friends. "With us everybody has learned how to use, how to keep under, and how to appreciate Bermuda grass," wrote a farmer there in 1848. "True, it is like some of our more valuable negroes —has no respect for lazy white folks . . . ."[17] Andrew J. Lane of Granite Hill said: "Had our ancestors, before removing from the worn out plantations of Middle Georgia—from whose fine productive soil many of them skined [sic] out a fortune—planted them out in Bermuda Grass, instead of leaving them in deep gullies and red hills, they would have, in great measure compensated those who had to come after them . . . ."[18]

Bermuda grass was slow to arrive in the northern part of the state, largely because farmers who moved there from Middle Georgia were careful to bring none of it with them. Aversion to grass in that area, however, was never as conspicuous as it was in the Cotton Belt. With large areas of limestone soil, this section gave promise of developing a thriving livestock industry. In the fertile valley areas of this region, there appeared a number of substantial intelligent farmers who recognized the livestock possibilities of the area they inhabited, and they developed a philosophy of agriculture based upon grass and livestock which crystallized into a practical system during the 1850's. The leading figure in this development was Charles Wallace Howard. During his travels in Europe he had been impressed with the settled agriculture, admiring particularly the livestock farms of the Scandinavians. Giving up the pastorate of the Huguenot Church at Charleston in 1852, Howard established a small home boarding school at "Springbank" in Cass County where he had acquired a large farm.

Here he started a crusade in behalf of grains, grasses, and livestock. As editor of the *South Countryman* and later of the *Southern Cultivator*, he acquired a large and respectable following.

With the persistence of a Cato, Howard propounded theories on grass farming which he had developed out of wide travel and extensive reading. He had observed that wherever improved agriculture existed there were always large numbers of livestock which produced an abundance of manure. It was this manure which was responsible for large crops, and wherever land produced such crops it was exceedingly valuable. Thus he attempted to adapt to a frontier society the old Danish maxim, "No cattle, no manure, no manure, no crops." In describing the pleasures and comforts of a grass farm he called attention to the fact that a cotton plantation required all of a farmer's time. The grass farm, he pointed out, gave leisure for reading, study, and the amenities of social life. "When we add . . . [the products of livestock] to our cotton, rice, and sugar," he said, "we shall, perhaps, live more independently than any other people in Christendom." He maintained that the low price of land in the South was not caused by slavery, climate, depressed agricultural prices, or even by the presence of cheap land in the West, but rather by the South's unbalanced and defective system of agriculture.

While he advocated slave labor, Howard deplored the fact that landed property had no value independent of the labor bestowed upon it. "The value of the negro is instantly affected by a change in the price of cotton," he said, "while the value of the land which grows the cotton, is comparatively unaffected. It is an extraordinary anomaly that perishable labor should take precedence over imperishable land." He cited the case of a Belgian farmer who sold his land in Europe for five hundred dollars an acre and bought rich river bottom land in Floyd County at twenty dollars. Having been informed that Georgia land would not grow grass and clover, the Belgian began growing the traditional crops of cotton and corn. He discovered to his dismay that he had enjoyed greater prosperity on expensive land under the European system than he had enjoyed in Floyd County under the prevailing system.[19]

Howard argued that raising stock extensively on grains, as was often done in the South, was too expensive and impracticable. He pointed out that the South had the advantage over the North in growing grass crops during all seasons of the year, an advantage which was enhanced by a saving in the cost of harvesting and storage. "The cost of one Pennsylvania stone barn, would lay down a large Southern plantation to grass upon which cattle sheep and horses may graze during the winter without the trouble and cost of making it into hay," he stated.[20]

In urging farmers to grow more grass, he cautioned them to prepare the land properly and devote the same attention to it as they would to a crop of cotton. He believed that the inaccessible creek and river bottoms of southern Georgia, where native grasses had not been exterminated by cultivation and by excessive grazing, might produce a desirable native grass. "Whenever a tuft of unknown grass, green in winter, is found, let it be protected or removed and its habits noticed," he urged. "No one can foresee what results may follow an intelligent examination into the native winter grasses of Georgia." While entertaining the hope for a native grass he conducted experiments with grasses imported from Europe, among which were vetch, rye grass, lupin, and sanfoin.[21]

Howard considered broomsedge as the greatest enemy to permanent pastures. While observing its abundant growth in old fields, he noted that it did not grow around a burned stump. This led him to conclude that broom sedge land was deficient in potash and "ammonical manures." In order to restore to fertility large areas of Middle Georgia he believed it was necessary to convert these lands to Bermuda grass and sheep walks.[22]

Howard believed that Lucerne, or alfalfa, held great promise for Georgia conditions. This hay crop had long been cultivated by a few farmers in the eastern states where it was introduced in colonial times, but it was found only rarely in Georgia before 1850. Shortly after this date it was brought to the West Coast from South America and reintroduced as Chilean clover. It was not widely popular in Georgia at any time before the Civil War, yet it was recommended by those who tested it as excellent food for cattle.[23] While it was easy to grow on the limestone soils of North Georgia, elsewhere it was found to require strong applications of bone dust and stable manure. Richard Peters estimated a cost of fifty dollars to prepare an acre for Lucerne.[24] Such a requirement was a serious drawback to its early popularity in the Cotton Belt.

During the Civil War Howard noticed that a field of Lucerne on his farm, which was completely destroyed by the grazing of Sherman's horses, produced the next year a very heavy crop of corn. This caused him to suspect the nitrogen-fixing qualities of the plant. These valuable qualities remained generally unknown for some time, however, and interest in the crop lagged. As late as

1880, Howard stated his belief that there were not more than ten acres of alfalfa on any farm in the South. With the practicability of Southern hay culture firmly established, he complained that Southern railroads were "still groaning under the weight of Northern hay."[25]

While Howard was developing a consistent philosophy on grass and livestock, another North Georgian, Richard Peters, was engaging in what was perhaps the most extensive series of experiments in grass and livestock farming to be found anywhere in the ante-bellum South. Born in 1810 at Germantown, Pennsylvania, of land-poor parents, Peters was the grandson of the famous Pennsylvanian for whom he was named. He received a limited formal education in engineering subjects in his native state, and then came South in 1835 to work on the construction of the Georgia Railroad. Two years later he became the superintendent of the road at an attractive salary. A keen business man, he used his savings to buy cheap land along the road's right-of-way, and operated sawmills. When the village of Marthasville showed promise of growing into the metropolis of Atlanta, he moved to that place, in 1846, and bought land in the vicinity. The increment values of his land investments made him one of the wealthiest men in North Georgia, and he engaged in other enterprises from which he realized additional wealth. Most of his permanent investments in farm land were in Gordon and Cass counties. He became interested in this region during a deer hunt when he noted the similarity in geological structure and soil formation between it and the land he had known in Pennsylvania.[26] Peters' wealth was matched by his interest and enthusiasm for agricultural experimentation, and he used his resources unstintingly. Martin W. Philips of Mississippi said in 1856 that Richard Peters was doing more for his region than any other living Southerner.

With a flair for intelligent planning, Peters began his livestock industry in a quest for grasses. He is said to have tried as many as forty varieties of European grass seed in one year, on the theory that Northern grasses would not thrive in Georgia. He discovered, however, that the better known grasses already being grown in the North and West, such as orchard grass, red clover, alfalfa, and bluegrass, were the best for North Georgia.[27] In 1861 he had in woodland pastures and meadows about three hundred acres of alfalfa, red and white clovers, wild oat grass, and others. Peters believed that red clover was the ideal grass crop for most purposes. Unlike alfalfa, it could be grazed rather than harvested for hay, and it required land of only average fertility. It was also highly adaptable to a rotation system.[28]

Peters began experiments with livestock in 1847, at which time he was one of the few breeders in the Lower South. Many facts which he determined by costly trial became axiomatic during his lifetime. It was his policy to spare no cost in securing the best animals for testing, and he claimed to have spent fifty thousand dollars for different breeds of animals which after experiment proved unsuitable to Georgia conditions. His most emphatic advice to farmers was to improve the breed of their livestock, and he insisted that it cost less to maintain a good animal than a poor one. "It is hard to calculate how much good a fine, vigorous Jersey bull can do in a country neighborhood," he said.[29] By using the male offsprings of purebred male animals for beef, and crossing the female offspring with other purebred males, Peters claimed that any farmer could upgrade his herd with minimum cost. It is highly improbable that the Peters farm paid any dividends before the war, but numerous farms in North Georgia could trace some improvement in their stock directly to the Peters enterprise.[30]

Progress was slow in the improvement of dairy herds. Throughout the period travelers uniformly condemned the milk and butter found on the tables of Piedmont planters. Although Durham and Hereford cattle had been imported from England at an early date, they were found to be a disappointment largely because of their susceptibility to fatal diseases.[31] The upgrading of native stock appeared to offer the only solution. Daniel Lee analyzed the situation in the following manner:

> It appears to us that all we need in the Southern States, to enable us to have beef, milk, butter, and cheese as good as man can desire, is . . . judicious selections from our own stock—then good pasture in summer—plenty of hay and good shelter in winter—and gentle treatment at all times. With means of this sort in abundance, our native stock will be found . . . to answer all our purposes; especially if our servants, and very many masters, too, can be induced to remember that there is some little difference between a cow and a mule . . . . What else than bluejohn in the very highest state of perfection, can be expected from cows fed on shucks and exposed to wind and rain all winter—turned out to sedge-grass pastures in spring, and brought up every evening in a trot or gallop, by the little negroes and dogs, or a big negro on horseback.[32]

Charles W. Howard also advised the farmer not to depend upon

Negro labor for the operation of any kind of livestock enterprise, for it would bring about the ruin of any livestock establishment. "A careless negro will soon make the best cow unprofitable," he said. "Unless he has a stirring active wife," he suggested, "it will be best for him to let a dairy alone."

The best imported Durhams had cost Peters nine hundred dollars, and other English cattle were equally costly. The expense, combined with a susceptibility to disease, caused Peters finally to reject all European breeds. After investigating livestock breeds in more temperate latitudes, he turned his attention to the sacred bulls of India. Dr. John Bachman of Charleston suggested that he conduct an experiment with Brahman cattle, believing the breed to be ideally suited to the Georgia climate.[33]

Brahman cattle were found to thrive in North Georgia. Peters, after spending more than twenty thousand dollars on cattle experiments, decided that he finally had the answer to all his problems. The Brahman cow made excellent beef and some of them proved to be good milkers. One yielded thirty-six quarts of milk in one day and won premiums at state and county fairs. The milk, however, was found to lack butter-making qualities and the animals themselves had a distinct aversion to women milkers. When these characteristics became known the Brahman's original popularity was quickly deflated. By crossing with the Devon, Alderney, Durham, and Guernsey, Peters hoped to soften the Brahman's temper and to improve the quality of the milk. The crosses proved to possess physical vigor and to be immune to diseases, the males making excellent work oxen. However, the thin milk and ugly temper of the females remained. "You can't sell a vicious cow," said Peters. "The women get against them and that ends them." On this note Peters ended the Brahman experiment after a long and costly effort to establish the animals as a permanent addition to Southern livestock.[34]

At the same time that Peters was working with improved cattle breeds in northwestern Georgia, others in the eastern mountain region near Clarksville were pioneering in the techniques of commercial dairying and cheesemaking. In 1822, Edward Williams came to Georgia from Massachusetts and bought a large part of Nacoochee Valley, containing about two thousand acres of level land. In 1848 he established a dairy of nearly one hundred cows on the top of Tray Mountain, where a young Vermonter, Joseph Hubbard, was engaged to superintend the work, using Negroes to assist in the enterprise. Shallow tin pans, placed in water flowing from a springhouse, provided the cooling medium for what perhaps was the first commercial dairy and cheese manufactory in Georgia.[85] It is not certain that Williams' plans included visitors who might observe his process, for Tray Mountain was almost inaccessible. After announcing a satisfactory experiment, he moved his dairy to the valley, where he enlarged it and installed water-power to propel some simple machinery.

Williams' death in 1856 brought this enterprise to an end. George H. Waring was operating a small dairy near Clarksville on the Souque River on the eve of the war, but little achievement was made in commercial dairying in Georgia throughout the period.[86] Charles W. Howard in 1862 stated that there was no dairy farm anywhere in Georgia. He called attention to dependence upon the North for milk products, which source was then cut off by the war. Not only had cities such as Savannah and Augusta been dependent upon Northern supply, but interior towns and villages as well. Butter was selling at fifty cents a pound, and it could not be obtained in sufficient quantities even at this price.[87]

The use of cattle for draft purposes was widespread in Georgia throughout the entire ante-bellum period. While work oxen were widely used by the poorer whites of all regions, they were also employed by many large planters in the Cotton Belt.[88] Work gearing for oxen tended to vary somewhat from one locality to another. In the Cotton Belt they worked in cotton and leather collars as well as in wooden yokes, the latter being universally used in the more backward regions. Some were driven by a bridle and bit and others by a head stall. In a few instances heifers and milk cows were used by subsistence farmers for light work in the fields. While there was a tendency to replace oxen with mules on the larger plantations after 1840, the total number of the former increased slightly in Georgia during the last decade before the Civil War.

The mule's capacity for longevity and hard work under conditions of ill-treatment and poor management had long been recognized, but since the supply came principally from Kentucky and Tennessee their use was limited. Only a small degree of local breeding was attempted before 1850.[89] In that year the ratio of mules to horses in Georgia was nearly one to three. By 1860 there were nearly as many mules as horses in the livestock population of the state, the latter being outnumbered by mules in thirty-one counties. Mules were concentrated in the cotton-growing counties, with Southwest Georgia having the largest proportion. North Georgia contained the larger proportion of horses, while the Wiregrass and Pine Barrens regions had the largest percentage of oxen. Commenting upon his decision to abandon work oxen in favor of mules, John Pitts referred to oxen as "dead weight" on any farm. "Steers want action," he continued, "and they eat up and destroy all the oats and fodder a man can raise . . . and have nothing to show for it but ragged hips that stick out from their backs like a lady's bustle."

Georgia at no time produced a significant part of its own requirements in mules. At the state fair in 1853, L. C. Warren of Jefferson exhibited a small number which were grown in the state. In 1855 the best grade of Kentucky and Tennessee mules sold for three hundred dollars a pair.[40] During that year it was estimated that four thousand of the animals were sold to Georgia planters. There was considerable speculation in the mule market. The difficulty of judging a mule's age led many unscrupulous dealers to buy up old and spent animals which they fattened up and re-sold at high prices. As a result, planters learned to buy only unbroken mules for fear of being cheated.

The breeding of horses on a commercial scale also received scant attention in Georgia. Richard Peters very early found such an enterprise unfeasible.[41] With the growing demand for mules that accompanied the expansion of cotton-growing in the Southwest, brood mares came into increasing demand in Tennessee for breeding of the hybrid stock. This demand was largely responsible for the scarcity of horses for all other purposes. In 1859 it was pointed out that horses which sold for seventy-five or eighty dollars twenty years earlier by this time were bringing as much as five hundred dollars. The price ranged upward to seven hundred dollars for a good carriage horse. Three years later, in 1862, the difficulty of raising a cavalry company in Bartow County was attributed to the excessive breeding of mules.[42]

Dire predictions were made during the 1850's concerning the destiny of horse flesh in the Lower South. "The best stallions [are] poorly patronized," said one, "and mares of finest form, the purest strain and most brilliant escutcheon, [are] basely prostituted to the forced and ignoble embrace of the assinine ravisher," lamented one lover of horses. It was in this period that buggies became popular, being substituted for horseback and to some extent for heavier carriages. The scarcity of horses produced in many instances the spectacle of a mule-drawn carriage. This so dismayed one Georgian that he began a crusade to restore the horse to his former position of dignity and aristocratic standing. "I have looked with amazement on the number of mules now used in buggys and pleasure carriages," said he, "and when I have seen a fine carriage filled with pretty girls and the farmer's wife, drawn through the streets by a pair of braying mules, I have regarded him as fit subject for a commission of lunacy."[43]

However, sentiments such as this probably helped to stimulate only a casual interest in horse breeding in ante-bellum Georgia. A prize-winning stud owned by W. A. Lenoir of Tennessee stood the season of 1853 at White Sulphur Springs in Meriwether County where horse growers of that section took the opportunity of improving their stock. Charles W. Howard, in an effort to encourage Georgia-bred horses, pointed out that the Morgan horse, being an all-purpose animal, rendered unnecessary the race course as a stimulant to breeding. Some interest was shown in the Morgan horse just before the outbreak of the Civil War. George H. Waring possessed on his thousand-acre grain and grass farm on the Souque River in 1860 what was thought to be the finest herd of Morgans in the state.[44]

Like horses and mules, sheep appeared in Georgia at an early date, although in limited numbers. In 1839 there were in the five southernmost states of the Union upwards of six hundred thousand sheep, nearly half of which were in Georgia. These sheep were poor in quality, however, and they roved the countryside untended. They found ample subsistence on such herbage as Georgia afforded, but encountered formidable enemies in the wolf and the dog.[45]

Sheep husbandry began to attract attention in Georgia about 1845.[46] Interest arose partly from the need to render economically productive the thousands of acres of Pine Barren and mountain lands which were not adapted to the staple crops. "If sheep husbandry is made a source of immense profit in countries where land costs thirty to fifty dollars an acre, and the climate requires them to be fed eight months a year . . . how much more available are our pineries at ten to forty cents per acre, with a genial climate and a soil of perpetual herbage?" wrote Jethro Jones of Burke County in 1845.[47] Henry S. Randall, an official of the New York Agricultural Society, produced facts to show that wool could be produced much cheaper in Georgia than in New York. He believed that growers in the latter state would stand no chance in competition with their Southern rivals.[48] He also believed that "keeping Alpacas on the mountain ranges . . . would yield the planters large profits, and compensate them for the low price of cotton."

Thomas Spalding suggested that the state spend a thousand dollars to secure ten Peruvian sheep for trial in the mountain area. It was stated that a pound of fine wool could be produced in Georgia as cheaply as three pounds of cotton and would yield a substantially larger profit. Henry J. Schley of Waynesboro reported that, in 1853, forty-two common ewes yielded an average of four pounds of wool apiece, which sold at thirty cents a pound; in addition, the ewes bore an increase over the previous year of forty-seven lambs. The average clip for common sheep, however, was nearer two pounds, and the price was seldom as high as thirty cents. The merino, a popular improved breed, averaged about five pounds.[49] There were many practical obstacles to a thriving sheep industry, however. James W. Watts of Cassville, who began breeding sheep in 1848, emphasized such problems as "dogs, thieving negroes, indolence, and ignorance." He suggested Spanish sheep dogs as a solution for the first two, and education for the others.[50]

The failure of the legislature to encourage the sheep industry when urged repeatedly to do so through enacting a simple tax measure on dogs is one illustration of that body's indifference to basic agricultural problems. The crusade against sheep-killing dogs was led by Henry J. Schley, one of Georgia's leading sheep enthusiasts. By 1840 wolves had ceased to be a menace to Georgia woolgrowers except in the more sparsely settled regions, yet dogs were so numerous as to be a greater hazard to the industry in the Piedmont than were wolves in the mountains and swamplands.[51] A New York agricultural editor stated in 1848 that he saw more dogs in Georgia than goats and sheep together. He attributed the great number of dogs to the presence of Negroes, each of whom he claimed was "ambitious to be master of dog, as he cannot be of himself."[52] Finally, in 1862, Schley succeeded in obtaining passage of a resolution providing for taking a census of dogs by having taxpayers state in their returns the number of the animals on their premises.

The success of Schley's oblique attack on the dog problem was attributed to an increased interest in food production following the outbreak of the war in 1861. Taking advantage of the anti-dog sentiment, wool-growers advocated the use of dog leather for shoes, adopting the slogan, "less dogs, more mutton and more shoes for soldiers." The editor of the *Augusta Chronicle* stated that there were enough dogs in the state to shoe every soldier in the Confederate army, and they consumed food which the people needed for their own subsistence. Schley's final efforts to secure passage of a dog-licensing law during the war were met by formidable opposition. The measure was defeated, as one complained, "by demagogical appeals about the poor soldier's dog, and the widow's dog." After a long and useful tenure in the legislature Schley was defeated for re-election in 1863.[53] Daniel Lee observed that "Cotton lands must be far more worn and washed, and less abundant before any planters of the South will abandon cotton culture and dog husbandry for wool growing."[54]

The scientific breeding of sheep was practiced in Georgia by only a few individuals before 1860. Among these were Richard Peters and James Watts who began the enterprise about 1847. Peters secured Southdowns, and later the Cotswold, Oxford downs, Leicester, and finally French merinos. Watts experimented with more varieties than Peters, among them the African broadtail, bred from importations from the African coast made by Richard Singleton of South Carolina. Both Watts and Peters turned eventually to the Spanish merinos as superior to all other breeds for the Georgia climate. Peters recommended them as making excellent crosses with native sheep, giving weight and fineness to the fleece and immunity to diseases commonly attacking the imported breeds.[55]

There was a paucity of merino flocks in Georgia at the end of the period, however. In 1859 there were only three small flocks in Cass County, which was in the center of the principal sheep region. One of the flocks was owned by Watts and the other by Charles W. Howard. In contrast, Georgia's native sheep population had almost doubled since 1840. Peters gave up the sheep industry before the Civil War, apparently because of the ravages of predatory dogs. He continued the industry on his ranch in New Mexico, experimenting on crossing Mexican ewes with rams from his Georgia flock.[56]

Despite his abandonment of sheep, Peters came to be recognized as the state's highest authority on wool-growing. He never abandoned the idea that under intelligent management, the industry could be made highly profitable, particularly in the wiregrass region of South Georgia.[57] There Bermuda could be combined with wiregrass for spring and summer grazing and shepherds as well as dogs could be utilized profitably. A serious drawback to the industry in ante-bellum Georgia, however, was the lack of a market for mutton, for Southerners, both black and white, had an antipathy toward goat flesh as a food.[58]

The abandonment of sheep breeding by Peters was the harbinger of his experiment with Angora goats, which animals he found more profitable. The breeding of the Angora in the South was suggested as early as 1828 but the first importations were made twenty years later by James B. Davis of South Carolina, at the time a consul in Turkey. Peters obtained his original stock from the Davis importation and carried to a final conclusion the work begun by the South Carolinian.[59]

In 1854 Peters bought from Davis his entire stock of full-blooded female goats and some of the males. In addition he made two large importations of selected animals from Turkey. He had difficulty in obtaining the best strains of Asiatic wool-bearing goats because of the closely guarded monopoly of the native owners. He found that his imported bucks had been rendered sterile before shipping, by the use of a hot iron rod. Finally he sent a man of stern qualities all the way from North Georgia to Turkey and obtained a good buck. Peters discovered that Angora goats could live frugally on scanty grasslands, and consumed herbage which other livestock left ungrazed. Unlike sheep, they did not need to be guarded carefully, for they fought off attacks of predatory dogs. They grazed over a wider range than sheep, required less water, and they could also be taught to come home regularly at night.[60] "You can count on a flock doubling every year while sheep . . . average eighty percent increase," wrote Peters. By crossing the Angoras with common goats he found the flesh of the crosses "superior to most mutton, tender and delicious, making them a desirable acquisition to our food producing animals."[61]

Peters set out to disprove the old adage that one could not go to a goat house for wool. One of his most valuable discoveries was that fleece-bearing animals could be obtained rapidly by crossing Angora bucks with native ewes. The fifth cross, commonly called a full-blood, yielded a fleece comparable in quality to much of the mohair imported from Asia. The wool of the Angora averaged three pounds per head and brought from thirty to sixty cents a pound. Before the Civil War, all Angora wool produced in Georgia was marketed in England.[62]

Among the goats imported were some with a greasy fleece like that of merino sheep. This fleece required the extra trouble of washing and it proved no finer and no heavier than that of the original Davis goats. Because of the practice of the Asiatic breeders, who did not pen their animals and hence did no selective breeding, it was discovered that no really pure-bred Angoras existed. Mated in the haphazard manner of range animals, some of the imported goats were parti-colored, and some had heavy manes along the neck and backbone.

Peters claimed that it took him twenty years of breeding and selection to establish a perfect animal and to eliminate the coarse mane. He found that only the male improved the stock, the characteristics of the female being recessive.[63] He thought that no breeder, however experienced, could select a so-called "full blood" from a flock of "thoroughbred" Angoras, "but the get of the full blood will invariably enable any intelligent shepherd to detect the fraud upon the thoroughbred," he said. Consequently he came to rely altogether upon thoroughbred bucks. "The progeny of the so-called 'full-blood' bucks varies greatly, and the upward progress is by no means satisfactory," he continued. One of Peters' bucks, "Billy Atlanta," lived to be ten years old and sired about two thousand kids. In 1875 it was estimated that his blood coursed the veins of more than half the Angora flocks in the southern United States.[64]

Another problem which Peters encountered was the tendency of the Angoras to shed their fleece, the unsheared goats losing their overcoat of mohair in March or April. Their winter coating then grew slowly until about July after which it increased rapidly until January when it attained its full length, averaging about nine inches. Peters found that if the goats were taken off the range and placed in pastures free from briars and underbrush around the end of January, their fleece was kept in good condition.[65]

The production of mohair, however, was never contemplated by Peters. His purpose rather was to improve the animals and to provide growers with good breeding stock. In this he could claim a high degree of success, for many famous flocks in other states branched from the Peters herd in Gordon County. In 1858, William Haupt, of Hayes County, Texas, secured eight of Peters' goats and began the famous Haupt flocks of western Texas. The last through train from Atlanta to St. Louis before the Civil War carried two young bucks destined for William M. Landrum of San Joaquin Valley in California. They went by riverboat to Fort Leavenworth and thence on foot with a wagon train and arrived in time to win a cup at the California State Fair in September. These goats, and subsequent acquisitions from Peters' flock, laid the foundation for the industry in California. After the war the industry was started by Landrum in several other Western states. In 1872 Brigham Young secured a small flock which began the Angora industry among the Mormons of Utah.[66]

When Sherman's army entered North Georgia in 1864, Peters drove his flock to the Florida line near Quitman and kept it there until 1866. It was this herd which gave him courage after the war to rebuild his farm which had been wholly devastated by Sherman's invasion. His record book shows that, between the close of the war and his death in 1889, Peters shipped his animals to numerous breeders in various parts of the Union, but principally in Texas, California, New Mexico, Tennessee, Kentucky, South Carolina, North Carolina, Mississippi, and Florida.[67]

In 1897 the last of the famous Peters flock was sold to J. R. Standley of Platteville, Iowa, whence many found their way into Canada. The Angora had at last become unprofitable in Georgia. In 1900 there probably were not more than three hundred Angoras in the entire state, yet they still were being recommended by agricultural leaders for the mountain counties.[68]

A complete analysis of the post-bellum decline of the Angora industry in Georgia which Peters established so successfully is not within the scope of this study. The causes for the decline, however, have important implications bearing upon the general trend and characteristics of the livestock industry in ante-bellum Georgia. The excellent quality of Peters' stock had been well demonstrated, and won the praise of stock breeders and wool merchants in every part of the Union. In his old age Peters stated that his Angora experiment yielded him a greater profit than any other of his extensive livestock enterprises.[69]

Surprised that the industry had died out in Georgia, a leading Texas breeder after the war suggested as the cause the impoverishment of the state's farmers. Since cotton was high and offered ready cash, Southerners devoted their energies to that crop alone. Those who owned good animals sought an income from the immediate sale of the animals instead of improving the flocks and selling the fleece. The flocks and their increase therefore were sold in scattered pairs and shortly the stock became diffused with that of common goats. The upgrading of common goats to Mohair production required about five crosses taking at least six years. The only by-product of the first four crosses was the flesh of the goats, for which no market existed in the South. Sometimes breeders were forced to accept as little as fifty cents a head for these animals.

As a result, farmers were discouraged from breeding because they did not have the capital resources for holding the goats and grading them up to an acceptable level.[70]

The lack of a more general interest in sheep and goats in Georgia may have resulted in part from the traditional popularity of hogs among Georgia planters. The scarcity of range land in the Piedmont Cotton Belt after 1840 rendered hog raising more costly there than previously. Hand feeding with corn was a common practice, since farmers had not yet learned to grow special crops on which these animals could graze. Outside the Piedmont Cotton Belt hogs still were raised largely on open range where they foraged for acorns and roots, supplemented by a little corn during the winter. The result was a half-wild hog, but a hardy and vigorous one. While his flesh was inferior, he paid well for the little attention which he required. The simplicity of this method of production, and the availability of cheap pork from Tennessee and Ohio were factors which retarded an interest in improving the stock.

The difficulty of keeping the hogs on the range side of common fences and their ravages to growing crops caused some planters in the cotton area to put them under special fence long before it was necessary to enclose other livestock.[71] The hog pasture made it possible to improve the stock through breeding, and a few enterprising planters early acquired pedigreed boars. The early maturity of hogs brought quick results from selective breeding. The hog pasture also made it possible to prevent winter farrowing, a condition which left pigs in the spring retarded by cold weather and frugal feeding.

The most common animal seen in the South in 1840 was the "land piker," a breed of common hog which was already fast disappearing in the North. The Berkshire hog, whose importation began about 1830, was the first improved breed to excite comment in Georgia. Among the early purchasers was John Bonner of Greene County, who bought a shipment from a New York breeder about 1840, and two years later began importing them directly from England. His hogs inspired enthusiastic comment by persons who viewed them. "Col. Bonner deserves much at the hands of his brother planters in Georgia for the zeal and enterprise he has manifested in the improvement of our stock of hogs," wrote the editor of the *Southern Cultivator*.[72]

Bonner claimed that he could obtain better weights than New York breeders, averaging four hundred pounds in one year. His method, however, was a complete revolution from that generally practiced in Georgia. During the early part of the season he put his pigs in clover pastures and fed them with potatoes or small quantities of barley meal. When green apples began to fall he turned the pigs into the orchards. In the late summer he fed them on peas while the vines were still green. When winter began he fattened them quickly for slaughter, using corn, barley meal, and peas.[73]

The work required by such excellent care and feeding did not appeal to the average planter. Nor could Bonner's skill in handling the various maladies which beset these unacclimatized breeds often be matched.[74] John Pitts was among those who experimented with the Berkshires for several years before turning again to the native stock. His problems were typical of those encountered by inexperienced growers of unacclimatized stock. He wrote:

> I have fed them bountifully on all sorts of grains, grasses, peas, potatoes, fruits, vegetables, meal and slops, and I have fed them scantily; I have enclosed them in lots, and I have let them run at large; I have fed them by themselves, and I have fed them with other hogs; but in spite of my best personal efforts, I have lost at least thirty of them to one of my common stock . . . . They would die poor, and they would die fat; they were subject to all sorts of diseases, old and complicated, new and simple; they would take the mange, and they would become lousy; they would die suddenly and they would linger to death.
>
> . . . I was cheered by the reflection that there was one experiment more to be tried . . . . [When the next litter came] I altered all the boars, spayed all the sows, killed all the hermaphrodites and knocked the old hogs in the head . . . [Now I am convinced] that this breed will soon pass away, and the sooner the better.[75]

One of the better methods of growing hogs in Georgia at this time was that of V. M. Barnes of Winfield, whose theory was to give them a short and merry life. Using only common stock, he fed his animals on what food the plantation afforded, usually corn and peas. He slaughtered them in December, when they were about a year old and weighed from 140 to 190 pounds each. Thus he made what he called "slaughter-clean pork" at a cost of four cents a pound. A Houston County farmer, by feeding common hogs on peas, potatoes, peanuts, corn, and kitchen waste, supplemented by summer grazing, felt that he was well rewarded by a slaughtered product which averaged 163 pounds.[76]

While lacking the usual vigor associated with native hogs and causing much initial disappointment, the Berkshire in Georgia was not a complete failure. At the fair in Hancock County in 1843, Bonner's hogs won five of the thirteen premiums, and his prize boar became famous in other states. Despite current skepticism, his persistence was rewarded when a committee on livestock at the State Agricultural Fair recommended the Berkshire as a satisfactory type for Georgia. Bonner's hogs sold readily to planters in adjoining states.[77]

Richard Peters' experiments with hogs in North Georgia were in a large measure a repetition of those of Northern growers. He began with white hogs developed in his native Chester County, Pennsylvania, but soon discarded them. They invariably took the mange, and the "buttermilk wash" used in Pennsylvania to prevent the malady was hardly feasible in Georgia. Peters later tried Poland Chinas, Suffolks, Neapolitans, Prince Alberts, Berkshires, and Jersey Reds, investing some $2,500 in them over a period of twelve years. About 1856 he imported a number of the Essex breed, which he found would produce meat at less cost per pound than any other hogs. The Essex was moderate in size and did not require unusual care. It was also suitable to the open range, a characteristic essential for the popularity of any hog in the South. Peters claimed that the Essex fattened to about three hundred pounds at one-third the cost of native animals.[78]

The best type of hog for Southern conditions was still a moot question in 1860. There were isolated examples of improved hogs on scattered plantations in Georgia, and the fact was established that success with improved hogs could never result from haphazard methods of breeding and care. There were pork shortages in various areas of the state at the outbreak of the Civil War.[79] On the eve of volunteering his services to the Confederate cause, Garland D. Harmon expressed chagrin and disappointment that planters everywhere in the South had not raised their own pork and other subsistence crops so that they might be independent of Northern supplies. "The voice of the cannon shakes the solid ground, and rivers of blood will be spilt," he prophesied. "Feed your people while they fight your battles. Drop off a few cotton bales, and raise corn and pork. Let us strike for independence, for home, for country."[80]

A more general interest was manifested in improved poultry. During the 1850's poultry enthusiasts included Richard Peters, George M. Battey, Dennis Redmond, Charles Collins, and J. V. Jones. More than seventy varieties of poultry were exhibited at the Macon fair in 1851. These included chickens of the red, black, and yellow Shanghais; Dorkings, black Polands, spangled Hamburgs, golden Top Knots, Bantams, Cochin Chinas, black Spanish, Malays, wild Indians, Bremen, Westphalia, and many varieties of Chinese geese. There were also numerous breeds of turkeys, peacocks, pigeons, quail, and ducks. The "guinea hen" made its appearance in Georgia at this time.[81]

The committee on livestock stated enthusiastically that it was doubtful if such an exhibition of poultry had ever been equaled anywhere in the United States. "The days of 'dung hills' are numbered," wrote an enthusiastic citizen of Atlanta in 1854. "No one except he be an age behind the times, would attempt the rearing and breeding of the 'common chicken' when their places can be supplied with those celebrated breeds from China and the East Indies." John M. Berrien of Rome reported that his Negroes raised 500 Shanghais for their own use in 1852, weighing from twelve to twenty-two pounds a pair. Dr. George M. Battey, who had a small farm on the east bank of the Oostanaula River, known as "Riverbank," was an enthusiastic poultry breeder. In addition to a great variety of chickens and geese, he also bred a large variety of pigeons, including fantails, carriers, drummers, pouters, capuchins, and tumblers.[82]

Richard Peters was an admirer of fine poultry and he experimented with various breeds. His more practical nature caused him to discard all varieties and breeds except Plymouth Rock chickens. These he considered ideal for the average farmer.[83] Many farmers increased their flocks and improved their breeds after 1850. By the outbreak of the Civil War there could be found in Georgia perhaps every variety of poultry known in the country. The center of the interest in poultry was in the upper Piedmont, on the nothern fringe of the Cotton Belt, somewhat in the same region as that of the broiler industry a century later.

X

Orchards and Vineyards

"HORTICULTURE, except in its lowest branches, is the science of a settled and permanent population, not the pursuit of a people struggling for bread and existence," spoke Bishop Stephen Elliott in a horticultural address at Macon in 1851. He pointed out that the largest and most valuable portion of the state had been settled only thirty years, and much of this had only recently been brought under cultivation. In calling attention to Georgia's proximity to a frontier society, Elliott added other factors which retarded horticultural development. Among them were the conservative resistance of planters to innovations, the absence of large nurseries for the propagation of trees, and the paucity of urban centers to provide markets for fruit.

Elliott deplored the necessity of Southerners' having to rely upon the horticultural literature of the Northern states and England, where conditions of climate and soil were not comparable with those in the South. "The countries whence we should have derived our horticultural maxims, Italy, Spain, and the countries bordering on the Levant have . . . been locked against us," he said. "And hitherto most of the experience . . . and observation of individuals has died with them, or has lain dormant . . . for lack of societies and journals through which . . . to communicate it to beginners."[1]

Horticulture in Georgia had its origin on the Georgia coast during the Colonial period. Before the severe freeze of February 1835, several fruits of a semi-tropical nature were still being grown on the coast. In that year John Couper of St. Simons stated that orange trees were killed which he believed had been growing there and on Jekyl Island for over a hundred years. Couper lost at this time eight hundred bearing olive trees and also some date trees.[2]

In the 1840's a revived interest in horticulture appeared in the Piedmont, where soil and climate called for a new type of horticultural emphasis. In 1831 a New Englander, Malthus A. Ward, was appointed Professor of Natural History and Botany in the University of Georgia, where, during the two decades which followed, he introduced and disseminated new fruits and flowers throughout the state. His work demonstrated the adaptability of Middle Georgia to the cultivation of many fruits. He contributed freely, but usually anonymously, to horticultural and scientific journals. He established a botanical garden on the University of Georgia campus in 1833 with plants he collected from all parts of the world.[3] Solon Robinson was among those who admired the romantic beauty of this garden. "An expenditure of three or four thousand dollars instead of the scanty pittance doled out to the gardener," he stated, "would make this garden a place for the Athenians to be proud of." The fame of the garden exceeded the support it received by university officials, who, shortly before the Civil War, allowed it to become abandoned.[4]

Throughout the late ante-bellum period Athens remained the center of the new horticultural movement in Georgia. In addition to Ward, the state's early horticultural enthusiasts included James Camak, editor of the *Southern Cultivator* from 1845 to 1847. Around his mansion he planted a variety of fruits to demonstrate their adaptability to the region, and he emphasized fruit-growing in the columns of the journal which he edited. Daniel Lee, who succeeded Camak as editor, continued the emphasis on fruits.[5]

The peach was the favorite fruit in the Georgia Piedmont throughout the late ante-bellum period, in many cases being the only fruit produced on farms and plantations. While Northern fruit growers used budded trees and preferred the freestone varieties, Southerners preferred clingstone seedlings.[6] Following directions found in a Northern journal, a Hancock County farmer in 1843 claimed to have produced the only grafted trees ever seen in that vicinity.[7]

While peaches were grown successfully in Georgia during the Colonial period, by 1800 a few large orchards existed in Wilkes County and other upcountry areas, some farmers having as many as five thousand trees. Since the fruit was highly perishable, the large orchards were planted to provide food for hogs and also for making brandy.[8] In the course of time peaches came to be neglected and many orchards gave way to cotton fields. A writer in 1845 complained that peaches had become comparatively worthless because of neglect of the trees. "Even the brandy that is made of it, compared with thirty years ago," he said, "is often little better than blue ruin or cockle burr whiskey."[9]

The prevailing method of pruning peach trees was to cut off the lower branches so that those remaining would be out of reach of grazing animals. Also this type of pruning permitted plowing close to the trunk. However, it caused many trees to become partly killed on the southwest side by the scorching afternoon sun. The trees were shaped more often with an axe than a pruning-hook, the hacked-off stumps sometimes protruding six or eight inches. "The tops of fruit trees are seldom skillfully shortened in, and properly thinned," observed Daniel Lee in 1848. "Whilst some are allowed to make too much wood, and over-long branches, others . . . bring to maturity a [sic] excess of young fruit."[10] Jarvis Van Buren in a trip through Middle Georgia in 1850 declared that he never saw one tree properly trimmed and trained in the umbrella fashion. He recommended allowing the limbs of the peach tree to grow from the trunk one and a half feet above the ground, and shortening them about half their growth annually so as to keep a new growth of fruit-bearing wood in the interior of the tree rather than at the end of the limbs. He cautioned against plowing nearer the tree than the spread of the limbs, so as not to injure the roots. Cultivation was limited to throwing furrows next to the trees, while rye, oats, or wheat was planted among them.[11]

Much of the poor fruit in the South resulted from failure to select good trees for planting and from planting of uncertified stock. Peach trees purchased in the North were said to be more successful than apple trees from that region, but Southern farmers had limited success with both fruits. A Hancock farmer stated in 1854 that thirty thousand dollars had been wasted in that county on worthless Northern trees. Similar views were voiced by planters in other states.[12] The yellows and curl, unknown to native peaches, sometimes attacked the less hardy Northern varieties, although they generally recovered. Damage from the peach borer and the curculio was increasing in the 1850's.[13]

With the declining price of cotton after 1840 peaches rapidly increased in popularity with the farmer. A few boxes of inferior fruit began to appear on the markets of Atlanta, most of which came from Morgan, Walton, and Greene counties. The expansion of rail transportation and the prospects for outside markets were important factors in this revival. An editorial in the *Southern Cultivator* in 1845 suggested that farmers near the railroads leading to Savannah, Charleston, Macon, and Montgomery use their poor sandy land, considered good for nothing else, in growing peaches for the supply of those markets.[14] Sir Charles Lyell noted that at Parramore's Hill, in Middle Georgia, a planter had decided against moving to Texas because he expected to reap a rich harvest from a thriving plantation of peach and nectarine trees just then coming into full bearing.[15] George W. Fish of Macon expressed satisfaction at seeing a renewed interest in the horticultural arts in Georgia and he called the attention of Northern nurserymen to the scarcity of commercial nurseries in the South, suggesting Macon as the logical place for the location of such an establishment. "I will also remark that on account of our Southern climate, vegetation grows so rapidly that the nurseryman would be rewarded for his labor in grafting, budding, etc., in half the time that is required at the North," he said.[16]

It was perhaps in response to invitations similar to this that Robert Nelson came to Macon the following year. Nelson was a political refugee from Denmark where he had been attainted for treason and had had his fortune confiscated because of his liberal republican opinions. His father was said to have been one of the larger farmers and stockholders on the European continent. Nelson reputedly held a Master of Arts degree from a European university, and previous to 1848 he engaged in the nursery business in the North.[17] Despite the handicap of his foreign accent, he came to be respected for his excellent conversation and his horticultural skill.[18] In 1852 he acquired ninety-four acres of barren, sandy land near Macon on the west side of the Ocmulgee River, where he established Troup Hill Nursery.[19] "If horticultural experience of thirty eight years . . . should be of service to my adopted country, I shall feel happy in giving to the public the result of study and labor," he said.[20]

Nelson began the improvement of his land by planting it in corn with an application of ashes for fertilizer. "I was struck by the expression, 'worn out soil,' " he said later. "I did not know the meaning of it." He plowed and dug his soil to a depth of two feet. He scraped together nearly anything he could find, including leaves, twigs, weeds, blood, charcoal, ashes, scraps of leather, soap suds, animal bones, and stable manure, and from these materials he made a compost heap. With this compost he mixed surface soil scraped from fence corners and similar places. He sprinkled the heap with salt and lime, and turned the pile over at intervals.

Noah B. Cloud later visited Troup Hill Nursery and found the soil "rich and productive," growing every type of garden vegetable, fruit trees, and a variety of ornamental shrubbery and flowers. Nelson was invited to speak on fruit culture before the Southern Cotton Planters' Convention at Montgomery in 1853. This address won him a wide reputation, and his nursery's patronage was extended to Alabama and Mississippi. Troup Hill, however, was destined for a short life under its founder. In 1857 it was sold by the sheriff at public auction to satisfy claims of creditors. Meanwhile Nelson had edited the horticultural department of the *Georgia Citizen*, and had published a treatise on manures designed for Southern farmers. Despite the failure of Troup Hill, its founder's integrity apparently remained intact, and he continued to enjoy the confidence of agricultural leaders.

After losing his nursery, Nelson moved to Augusta where for two years he assisted Dennis Redmond in the management of Fruitland Nursery. When Charles A. Peabody severed his connection with the *American Cotton Planter and Soil of the South,* in 1858, Nelson succeeded him as the horticultural editor of that journal. He then moved to Montgomery, where he again engaged in the nursery business. "[His] name and reputation as an educated and experienced Horticulturist, are well known in all this country among fruit growers, as familiar 'house hold words,' " wrote Editor Cloud in announcing his arrival at Montgomery. Finally, in 1862, Nelson became an assistant in the editorial department of the *Southern Cultivator*.[21]

Nelson's skill in propagating fruit, and his encouragement of commercial peach growing in Middle Georgia, contributed greatly to the stimulation of this enterprise in the decade before the Civil War. At the same time that he was criticizing Northern nursery stock and its peddlers, he discounted the prevailing notion that better peaches were grown from the seed. He insisted on grafting from parent stock in order to produce uniform fruit. When he located at Macon in 1848 he was told that he had come to the land of peaches, and that he would see peaches such as he never saw before. "And it was true," he related, "for never had I seen such an abundance of mean, dry, hog peaches as those that abounded here."[22]

Nelson believed that the South had every possible advantage

over the North in commercial peach growing, and that the Southeast eventually would supply the Northern markets. "An immense wealth is opened for us in those markets, which are never . . . overstocked before the month of August," he said, "and yet how very few persons are ready to profit by the great advantages of which we are in possession!" He demonstrated his point by sending a box of peaches from Macon to New York in early July 1853, where such peaches sold at that season at fifty cents each. The local market at Macon for the common variety was twenty-five cents a bushel.[23]

Nelson observed that quantities of Northern peaches were brought to Southern ports from New York and Philadelphia during September and October, after the season for local peaches had expired. "Why don't we raise such peaches at home?" he asked. "Why don't we make the tide roll back by sending our early varieties to Northern markets?" He pointed out the advantage of a longer season in Georgia, which also resulted in a highly superior flavor.[24]

The commercial peach industry which appeared in Georgia in the decade following 1850 experienced a rapid and phenomenal growth. Georgia peaches which were exhibited at the New York State Fair in 1848 had their flavor and eating qualities greatly impaired by being packed with pulverized charcoal in hermetically sealed cans.[25] Three years later, however, Raphael J. Moses made an experimental shipment of peaches from Columbus to New York, packed in a champagne basket. They went by stage to Macon, then by rail to Savannah, and thence to New York by steamer. Moses received thirty dollars for the lot. Satisfied with the experiment, he expanded his orchard. After the railroad was completed to Columbus, he shipped extensively to New York, receiving very remunerative prices. In 1861 his sales had reached $7,500 for the year. He then had one hundred acres in bearing trees and an additional orchard of young trees which had not yet reached production.[26]

By 1860 there were many commercial peach orchards of varying size throughout the central part of Georgia, a farmer near Montezuma reporting an orchard of 40,000 trees. Edward Bancroft of Athens solved the problem of high perishability by shipping the fruit in self-sealing cans, disposing of several thousand dollars' worth annually.[27] The *Augusta Constitutionalist* in October 1858 reported nine cars of peaches in one day from Macon and Columbus destined for New York steamers. That year the *New York*

Daily Tribune reported that the peach had been found "a very profitable fruit for shipment to the North, and large orchards are cultivated on the railroads leading to Charleston, Savannah, and Norfolk for the supply of the New York market."[28] The range in price that season was from three to fifteen dollars per bushel. Fruit of large size expertly packed commanded the highest prices.[29]

The Civil War put an end temporarily to this expanding enterprise. Georgia peaches flooded the Richmond market after 1861, but the larger markets of the North were isolated. Moses built a canning factory and a drying house to preserve unmarketed fruit, but the end of the war found his orchards badly in need of renovation, and his undertaking never prospered afterwards.[30]

While Nelson, Moses, and others were making some valuable contributions to the embryonic peach industry in Georgia, Jarvis Van Buren of Habersham County was conducting a crusade to get farmers to grow better apples in the hills of northeastern Georgia. Born in Montgomery County, New York, in 1801, Van Buren came to the South at the age of thirty-nine. Like Richard Peters, he came to Georgia as a railroad employee, having been trained as a foundryman and having been employed in the first railroad machine shop in New York in 1831.[31] A man of broad interests and numerous talents, he soon abandoned railroad employment and bought ten acres of land in Habersham County and acquired a family of five slaves.[32] He pursued such avocations as building houses for planters in the vicinity of Clarksville, painting in oil and water colors, and practicing law on occasion. He established "Gloaming Nursery" near his home at Clarksville, one of the early apple nurseries in the Lower South. Coming from an important apple-growing section in Mohawk Valley, Van Buren early recognized the horticultural possibilities of his adopted section. He became a prolific contributor to agricultural journals, through which he sought to develop the apple industry in North Georgia and to direct the attention of Northern fruit growers to the region.[33] There were few issues of the *Southern Cultivator* in the 1850's that did not reflect his enthusiasm for the development of apple-growing in his mountain community.

He was an indefatigable collector of Southern seedling apples, although most of his collection came from the old Cherokee orchards in North Georgia and in the upper Piedmont. He made long journeys over mountains to secure grafts from trees which some backwoodsman recommended, believing that the native Southern apples needed only to be collected and improved to become equal to the best apples of New York.[34] "There are thousands of native seedlings scattered broadest [sic] over the South," he said, "that were they collected and tested, would prove superior to much that is imported under high sounding names."[35]

As late as 1845, when Andrew Jackson Downing's *Fruit and Fruit Trees of America*[36] went to press, no such thing was known as Southern fruits. None was mentioned in Downing's volume except the Columbia peach and the Father Abraham apple. Patrick Barry's *The Fruit Garden,* published in 1851, listed 133 varieties of apples, only four of which came from the South, among them the limber twig and the red June.[37] At this time the nomenclature of Georgia fruit was in a chaotic state. Complaining of this situation, a Georgian in 1853 wrote: "We have only one peach, I believe, that has a name, the Indian Peach; and one apple, the Horse Apple. And depend upon it, a name is as useful to a fruit as it is to a man. It will not make its way in the world without a good one."[38] Indeed every individual cultivator seemed to have a special name for his own fruit, and a single variety bore a legion of names, even in the same locality.[39]

Van Buren sought the organization of a system of Southern pomology, suggesting a committee "to name such native fruits as are nameless, and lop off some that we have a super-abundance of."[40] Charles A. Peabody suggested that the Southern Central Agricultural Society engage Van Buren to tour the country and to authorize him to name the fruit which deserved classification. As a result, he made full-size drawings of thirty specimens and colored them with remarkable fidelity, thus beginning a system of nomenclature for native fruits.[41]

Before the end of the period the Pomological Society of Georgia was organized and became a thriving organization. With the assistance of other able horticulturists, including William N. White, Prosper Jules Berckmans, and William H. Thurmond, this society carried forward with other fruits the kind of work begun by Van Buren in promoting the apple industry.

In naming the seedling apples, Van Buren chose many Georgia place names, indicating their origin. He found in an old Indian field in Habersham County a famous keeping apple which he named Mountain Belle. The Yahoola was another famous apple of Indian origin, found in Lumpkin County. The Buckingham apple was said to have come from the Cherokee Indians in Cass County. The Toccoa was found in the orchard of Jeremiah Taylor near the famous falls by that name in Habersham County, while the Tillaquah came from an original tree growing near Franklin, North Carolina.[42]

Two apple varieties acquired from the Cherokees were commemorated in lyric poems by Francis Orray Ticknor in his "Junaluskee" and "Nantahalee."[43] Van Buren named the latter from the Cherokee word meaning "Maiden's Bosom." The original Junaluskee tree was owned by the famous Cherokee Indian of that name who refused to part with the ground on which it was growing but who subsequently sold the tree for fifty dollars. Other apples which appeared on Van Buren's list in the 1850's were the Yates, Rhodes Orange, Taunton, Hamilton, and Oconee Greening, all of which came from seedlings originating in North Georgia.[44] The Shockley was said to have originated on the farm of Waddell Hall in Jackson County, while the Donehoo Cling and the Jackson Cling were found growing in Clarke County. Other North Georgia apples were the Green Mountain Pippin, Nicajack, Julien, Golden Rustic, and the Vandiver.[45]

After naming and classifying the seedling apples, Van Buren set out to convince skeptics that Georgia could produce a superior product. He took issue with Daniel Lee, who claimed that he had never seen a healthy apple tree in Georgia and who did not believe this fruit could be grown profitably in the Lower South. In 1854 Van Buren sent some Georgia apples to a Pennsylvania fair where they attracted unusual attention. "Taking all the good qualities of these Southern seedlings into consideration," said an observer, "I doubt very much if a selection of an equal number could be made in Lancaster County, or east of it at all comparable to them . . . ." Van Buren challenged any Northern grower to compete with him in exhibiting any twenty varieties of winter seedlings. "We do not claim to be a prophet or the son of a prophet," he said, "but the North must pay back the amount she has received from the South with interest for fruit and fruit trees purchased from her . . . ."[46] Others predicted the end of the South's dependence upon New York, New Jersey, and Pennsylvania for the apple.[47]

The new apple industry was centered in Habersham and Hall counties, but farmers elsewhere were paying more and more attention to the selection of good trees and were realizing that good apples cost no more to raise than poor ones. As enthusiasm spread, the "apple a day" slogan became popular, and physicians recommended more fruit for Negroes to keep them in good health. Many professional people in Georgia became horticultural writers, the most famous of whom perhaps was Francis Orray Ticknor, of Columbus. Under the name of "Torch Hill," through agricultural journals he gave sage advice on fruit growing. The combination of his passion for fruits and flowers with his talent for writing lyric poetry made him a well-known figure in the fruit-growing movement.[48]

A typical Georgia apple orchard in 1860 was that of Colonel Singleton Buckner near Milledgeville. Although some distance below the recognized apple belt in Georgia, Buckner's orchard of fifty acres comprised 7,000 trees, of which more than ninety per cent were the Shockley or Romanite. His trees were planted seventeen feet apart each way, a crop of pears being grown on the same ground. While the land was not of good quality, some trees produced as much as eight bushels annually and the fruit was marketed principally in Montgomery and Savannah, where it sold for five to seven dollars a barrel.[49]

Nurseries, which were unknown in many Georgia communities ten years earlier, were doing a thriving business in 1860, and Northern establishments were doing extensive advertising in Southern journals. Gloaming Nursery's advertising copy claimed the greatest variety of Southern seedling fruits in the United States. William H. Thurmond operated a large establishment in Atlanta known as Downing Hill. There were nurseries in many smaller communities, such as Covington, Spring Place, and White Plains.[50] In 1854 Richard Peters engaged William P. Robinson of Cincinnati to come to Atlanta and supervise his nursery on Fair Street. Peters claimed to have imported 40,000 plants in a single year, including roses and evergreens. Much of this stock came from France, and his propagations in turn found their way as far west as New Zealand. On one order he shipped 4,000 peach trees to California around Cape Horn.[51]

Peters conducted experiments on a fifty-acre orchard checked off into plats on which the growth and development of trees were carefully observed. Unsatisfactory trees were discarded and new scions were propagated from more promising stock. He popularized many old garden fruits, including a cherry early brought from France by William H. Crawford and once known in Middle Georgia as the May cherry. Peters sold it widely as the DeKalb cherry. The nursery firm of Harden and Peters had a prize-winning collection of pears on exhibit at the State Fair in 1860.[52]

Pear culture underwent considerable development in Georgia during the decade of the fifties. Van Buren said in 1851 that "not one person in a hundred, in Georgia, ever tasted a good, melting, juicy pear, nor can they imagine how luscious a fruit it is, until they do." In 1857 the *Southern Cultivator* announced that Louis E. Berckmans, the famous Belgian horticulturist, was planning to move a collection of 20,000 seedling pears from New Jersey to the South. Berckmans became associated with his son, Prosper Jules, in the nursery business at Augusta in 1859. Two years later this firm, known as Fruitland Nursery, probably was the largest establishment of its kind in the South.[53]

Unlike the pear tree, which arrived late in the ante-bellum period, the grape vine was indigenous to the South and grew profusely throughout the area. Early Georgia settlers described the native or wild grape as generally of two types. One was called the fox grape, having large dark round berries, which appeared singly or in small clusters. Because it tasted like the musk grape, it was referred to as the bull or bullace grape. The other native was called the cluster grape. It bore black clusters, the fruit being a small berry with a thick skin and a large stone, the taste resembling the Bordeaux of France. In the late ante-bellum period this smaller grape came to be called the fox grape, taking the original name given to the muscadine or bullace variety.[54] On St. Simons Island was found still a third type, described as resembling the common white grape of Europe, although having a different leaf. This may have been the scuppernong, which derived its name from a swamp in Washington County, North Carolina, where it first came into notice by Europeans.[55]

The vines of all these natives tended to grow high into trees so that the fruit was eaten by birds and wild animals and it was difficult to harvest in sufficiently large quantities for wine making. As a result, the native grapes were neglected by early settlers in favor of imported varieties from Italy, Portugal, France, and other countries. But the imports did not thrive and the little wine made from them proved inferior. It was described by one writer as "sad stuff, and bitter, rather the juice of the stalk than of the grape."[56] The more skilled vignerons in Colonial Georgia included a German servant named Rinck and a Portuguese Jew, one De Lyon. The latter made an attempt to graft Portuguese grapes to the native varieties, but nothing of significance came from his efforts.[57]

Wine remained scarce and expensive throughout the eighteenth century. About 1816 Thomas McCall of Laurens County began a systematic experimentation with grape culture and wine making.

He read widely on the subject and kept meticulous records. Repeating the experience of earlier growers, he soon discarded imported varieties and turned to native grapes. He claimed in 1823 that he was the only vine dresser in the state. "The people of Georgia . . . call me a visionary and other names," he wrote. "They prefer Auguandente, Whiskey, and the more execrable Peach Brandy . . . to the beverage of the gods."[58]

McCall's wine made from native grapes became famous in many communities. In 1826 he produced 860 gallons from a two-acre vineyard. A renewed interest in wine-making arose in a few Piedmont communities but only a few enterprising individuals remained long in the business.[59] Wine-making from native grapes apparently was mastered by only a few men, and they kept the secret largely to themselves. With their passing, between 1835 and 1850, the art became lost, but cordials of grape juice, sugar, and whiskey were not uncommon. James Camak, in encouraging a revival of vineyards and wine-making in 1846, spoke of "the horrible rot-gut stuff" at that time being sold in Georgia as foreign wine. He lamented the absence of "good Georgia wine" which once had been made by McCall, Iverson Harris, and others. The gravel ridges of North Georgia were cited as similar to the fine grape country of Spain. In that region were some Belgian settlers near Rome, among them Camille Le Hardy, who was experimenting with both foreign and native vines.[60]

No suitable market for wine existed in Georgia at this time. James Horsley of Upson County made large quantities of scuppernong wine in 1845, but the lack of a market so discouraged him that he did not feel justified in preparing another vintage. "But such is the prevailing bias for everything foreign, that consumers of wine in this State will scarcely be induced to prefer the domestic unless it were offered to them under some outlandish name or unintelligible brand," wrote a correspondent to the *Southern Cultivator*.[61] Yankees were accused of buying Southern wine, adulterating it in their cellars, sending it back to the South under their own labels, and selling it at a great profit.[62]

The wine industry in the South remained largely a home enterprise, for there was no market for the inferior product of the unskilled Southern vintners. There were other factors to retard an industry based on native grapes, the hardiest and most popular of which was the scuppernong. This grape was never subject to rot or mildew and was a prolific bearer, but the fruit did not ripen uniformly. Even when growing on the same vine, the grapes matured a few at a time, extending over a period of six weeks or longer. The fruit appeared singly or in small bunches of two to six grapes, each berry tending to drop off as it came to full ripeness, so that harvesting was a tedious and endless process. Vintage had to be done at several different periods of changing fall temperatures, and the varying degree of ripeness of the fruit rendered the final product difficult to standardize as to quality, flavor, and alcoholic content. The juice often required additions of sugar or alcoholic spirits. While a prolific bearer, the scuppernong did not lend itself to the usual methods of cultivation. When propagated from the seed its white fruit would produce a black or purple grape, frequently called a bullace*. The scuppernong did not strike readily from roots, and it sometimes bore only staminate flowers.[63]

The bullace grape, of which the muscadine was one, was a wild grape of some popularity. Although there was no fundamental botanical difference between the muscadine and the scuppernong, the former bore a purple fruit, somewhat musky in taste, with a thick skin and relatively little juice. The cluster grape, then known as the fox grape, or wild grape, was sour, and it bore a small and very dark berry, the wine from which was of poor quality. However, unlike the other two, it ripened evenly and appeared in large clusters. It did not shatter easily and could be harvested with little effort. Because of these qualities vintners considered fox grapes the most valuable grapes from which to develop more highly flavored offsprings.[64] However, the scions did not prove entirely satisfactory. "I have no faith that we will ever get a first class grape from the Fox family," said a Georgia vintner in 1859. "All will have more or less a nigger stink, thick pulp, and disposition to rot."[65] All of the wild grapes had still another disadvantage to the Piedmont farmer. They matured in late summer or early fall and the harvest and vintage period coincided with the rush season of the cotton harvest, when all hands were busily occupied. Another problem with wild grapes was the difficulty of making wine from them without the use of sugar. A wine cellar to control the temperature was indispensible, and a wine cellar was rare indeed in the Cotton Belt.[66]

Between the delicious scuppernong and the prolific but sour

---

*The bullace grape was a term used in England and in the Colonial period to refer to any one of several varieties of purple grapes. In Georgia, by the middle of the nineteenth century its use was confined to the large purple or black grape growing in the woods.

fox grapes, however, Georgians had an abundant variety of summer grapes for table use, all of which by 1860 had undergone considerable modification. Chief among these perhaps was the Warren or Warrenton grape, named for the town of Warrenton, where it was cultivated in the early part of the nineteenth century. It had a relatively small fruit, which grew in large bunches, and was a reliable bearer. The Devereaux grape was found in the woods near Sparta in 1820 by Samuel M. Devereaux, who cultivated it and distributed it among his neighbors in Hancock County. Devereaux kept a stagehouse, and travelers who tasted this grape popularized and disseminated it throughout a wide area.[67] The "Old House grape," sometimes known as the Harris grape, was found at a deserted house near Milledgeville by the father of Iverson L. Harris. Other grapes of Southern origin cultivated more or less extensively in Georgia included the Catawba, Isabella, Lenoir, and Herbemont.[68]

The origin of most of these grapes is obscured by a collection of mythical and conflicting stories, for at no time before 1860 was there a standard nomenclature for grapes, many single varieties sometimes being known by half a dozen names. The Warren grape, for example, appeared in Thomas McCall's records as the Hunt grape. McCall claimed that it was brought to Georgia before the Revolution by Henry Hunt, a kinsman of the Earl of Shelburne, and planted in St. Paul's Parish north of Augusta, whence it spread to Warren County. It was later known as the Monroe grape because it came into prominent notice during the presidency of James Monroe. Joel Crawford maintained that the grape was named for William Warren of Putnam County, who obtained it from his son-in-law, David Neal, who was cultivating it extensively near the village of Warrenton as early as 1805.[69] Nicholas Herbemont claimed that he propagated it from an old vine on the plantation of John Huger near Columbia, South Carolina. Prosper Jules Berckmans maintained that it was first found around 1800 by Harmon Perryman when clearing a piece of woodland on the west side of Rocky Comfort Creek in Warren County. Seeing the thrifty young vine running over a tree, and saving it, Perryman found that it bore excellent fruit. He brought it to the attention of others, many of whom thought it to be indigenous.[70] There were others still who claimed that it appeared first in Warren County, North Carolina, where it was brought from the Cape of Good Hope. Only one thing is certain. The

Warren grape was a highly popular grape and was propagated by all grape enthusiasts who tasted its excellent flavor.[71]

All of these problems were reflected in the proceedings of the Southern Vine Growers' Convention held at Aiken in 1860. That they understood these problems and made progress toward their solution is a tribute to the men who led the movement for the revival of grape culture in the South. Taking a prominent part in the proceedings at Aiken were Victor LaTaste, Dennis Redmond, Prosper Jules Berckmans, William Schley, and others of Augusta; Jarvis Van Buren of Clarksville; S. W. Printup and J. B. Hart of Union Point; John M. Couper of Savannah; and J. M. Nunez of Columbus.

The conviction was expressed at the convention that only native varieties could be relied upon for extensive planting and for use in making wine. It was pointed out that the same variety of grapes under varying conditions of soil and climate and in different stages of ripeness produced wine of uneven characteristics, and standardization was made difficult. To remedy the problem and that relating to nomenclature it was recommended that vintners adopt a method of labeling which would show first the name of the state in which the grapes were grown, followed by the name of the town, river, or locality where the wine was produced, and last the private name or brand of each manufacturer. In the hope of solving other problems the convention voted to invite to their next meeting all vine growers in the United States.[72]

The convention appeared to be divided on the question of wine adulteration. Some contended that domestic wine was lacking in body; to supply the deficiency, said Professor William Hume, "we must supply Alcohol, good pure Alcohol, not the filthy Brandy concocted of chemical substances and colored with logwood and cochineal." The opposition to alcoholic adulteration had already won the first round of this struggle when it was decided to hold the meeting in the Baptist Church of Aiken and to engage a minister to preface the proceedings with prayer. The minister before his prayer made the statement that he had no objection to asking divine aid in the manufacture of "the pure, unadulterated juice of the grape," but suggested a moral obligation to oppose adulteration with alcohol. Despite this clerical head wind, a resolution was offered declaring that "a moderate addition of brandy . . . or a concentration of grape sugar, by desication [sic] of the fruit, or other means, is both proper and judicious." After considerable debate this resolution was voted down.[73]

The excitement in Georgia over grapes and wine coincided with the appearance in Augusta, in 1848, of a German vine grower, Charles Axt. Born a Rhinelander, he had engaged in the wine industry in France and Germany from his early childhood, and he had spent some time in the grape region of Ohio before coming to Georgia. He thought that he had found in the rolling Piedmont hills north of Augusta an ideal combination of soil and climate for the cultivation of grapes. Augustans, however, assured him that the section would not produce grapes to such perfection as he anticipated. Axt was confident that with good management the Georgia vineyards would exceed in production those in his homeland. He secured the confidence of several planters who each agreed to put in a vineyard of one-fourth acre. Axt agreed to furnish the plants and dress the vines for three years at fifty dollars a year, after which time he promised a production of 350 gallons of wine. In the meantime he was to teach the planters the skill necessary to continue the enterprise.[74]

At first Axt made slow progress and met with much discouragement. He complained that planters had no conception of the business. In 1855, about the time he began to master the English language, his project began to prosper. He had acquired management of a number of vineyards throughout the northern half of Georgia as well as in Alabama and South Carolina, and he procured experienced vine dressers to aid him in his work. He acquired a farm near Crawfordville, where he established a five-acre vineyard. "We hope to live long enough to see the old 'washed' and 'gullied' hillsides of Georgia and the neighboring States, yielding tons of luscious Grapes and hogsheads of pure and invigorating wine," said the editor of the *Southern Cultivator* in discussing Axt's success as "an itinerant Grape Missionary."[75]

Axt's perseverance apparently was rewarded. At the Atlanta fair in 1855 his grapes and wine proved to be an outstanding sensation, both being pronounced the most superior ever seen in Georgia. Axt was awarded a silver cup in recognition of his work. However, vintners from other states soon voiced considerable skepticism concerning his claims, particularly that it was possible for him to produce a thousand gallons of wine from a single acre in the third year after planting.[76] Skepticism as to these claims became so widespread that the *Southern Cultivator,* which had been enthusiastic in reporting Axt's activities, began to exercise caution. When the editor admitted that the German's imperfect knowledge of the English language may have caused him to err in making his reports, Axt maintained that no fact had been overstated. He continued to insist that his statements were not exaggerated. He invited visitors to inspect his vineyards and proposed a meeting at Augusta of all vine growers in the country, to compare wine samples and to prove that "better wine [could] be made in the South than in any other part of the world."[77]

While his critics did not show any desire to compete with his wines, a self-appointed committee of vine growers did assemble at Dalton in the summer of 1856 to examine some of Axt's young vineyards. The Dalton vineyards consisted of nine and one-half acres of two-and-one-half-year-old vines belonging to James and E. W. Green and Lawrence Wilson. The committee reported an average of fifty-five bunches of grapes on each vine of the Catawba variety. One vine had one hundred and two bunches, with a single bunch weighing one and one-fourth pounds. There were sixteen hundred vines to the acre. Properly impressed, the committee rendered a glowing account of Axt's success as a grape culturist. Before they left Dalton they formed the Vine Growers' Association of Georgia, a short-lived organization destined to succumb to the pressures of the Civil War. "Certainly we have seen nothing of the grape kind that equalled these, either for richness of flavor or luxuriance of growth," said the editor of the *Atlanta Intelligencer* about Axt's Catawbas.[78]

In 1857 Axt's "Still Catawba" was becoming well known on many local markets. This wine was put in specially made hock bottles, well corked and attractively labeled. Axt offered to sell at reasonable prices the rooted cuttings from his vineyards at Crawfordville, Washington, Augusta, Atlanta, Dalton, Montgomery, and Abbeville. "My wine has stood the test of the best judges," he boasted; "it is now in market, and will rest on its own merits."[79]

The culmination of Axt's success was reached in 1859 when his Georgia wine of the 1857 vintage received wide acclaim at Cincinnati, and commanded a higher price than the Ohio wines. An observer wrote that "gentlemen [in Georgia] can scarcely be aware of their advantages in soil and climate, but, sooner or later, they will find it out."[80]

The 1857 vintage was perhaps the most famous wine that Georgia produced in the ante-bellum period. The following year in most of the state the vines were badly injured by late frost. In 1859 excessive rain caused damage to the fruit, resulting in an inferior vintage, although the scuppernong and a few of the hardier varieties escaped. The year 1860 was more favorable, but not equal to 1857 in either the quantity or the quality of the fruit.[81]

Despite the circumstances causing short crops, there was little abatement of enthusiasm for the new enterprise. In 1860 John Winn of Mallorysville was making plans for a vineyard of two hundred acres. Near Macon were many large vineyards, among them that of John M. Fields, containing five thousand bearing vines and several thousand rooted cuttings for the expansion of the vineyard. Raphael Moses of Columbus had an eighteen-acre vineyard of Catawbas under the management of a French vine dresser. Indeed the promising Georgia wine industry had attracted attention of the New York press.[82]

An observer at the meeting of the Georgia Pomological Society in 1860 noted that within the decade greater progess had been made in grape culture than in any other branch of horticulture, and the future of all horticultural enterprises appeared to be exceedingly bright. It had been demonstrated that many varieties of fruit could be grown with success in the South, and that Georgia products could find a profitable market in the Northern cities. North Georgia apples already were selling in local markets for as much as two dollars a bushel.[83] "No part of the United States has progressed more rapidly in Pomology than has the Southern States for the past eight years," wrote Van Buren in 1860. "The quantity exported will, next year . . . be enormous. Apples in the upper portions of Georgia, where cotton does not grow, are becoming a staple production; and even this year, when the crop is a failure . . . some forty to fifty thousand bushels of apples will be carried to market."[84]

Jarvis Van Buren thought that the greatest problem in the development of horticultural crops in Georgia was lack of transportation. He stated that five or six counties in northeastern Georgia could have sent to the seaboard in 1860 at least a hundred thousand bushels of apples and he thought that amount would be quadrupled in ten years if transportation facilities were made available. As it was, thousands of bushels of peaches and apples rotted on the ground each year. Van Buren also complained that "politics, cotton and corn" were still the all-absorbing topics when Southerners came together, and that the South's enemies were still supplying the region with what could be found in abundance locked up in the valleys and mountain slopes of northeastern Georgia. "With a climate and soil capable of producing every desired production, they turn to the frigid regions of the North for everything consumed in their families, and which a little effort could supply . . . at home," said this former New York foundryman, who had now become a thoroughgoing exponent of Southern agrarianism and economic independence. "Although our head is slightly seared by frosts of time," he continued, "yet we trust to see the 'sunny south' scattered over with tasteful cottages and the surroundings of thrifty orchards and glowing flower gardens . . . the external harbingers of high civilization, intellectuality and progress."[85]

XI

Gardens and Buildings

HORTICULTURAL activities after 1840 included an interest in the vegetable garden as well as in flowers, ornamental shrubs, and trees. The beautification of the homestead focused some attention on construction of better farm buildings and on the architectural design of the planter's residence. While the larger cities of Savannah and Augusta were early centers of a developing esthetic interest in horticulture, the Piedmont Cotton Belt was slow in acquiring such tastes. Under the traditions of the open range the upcountry farmer had to construct strong, costly fences around his garden, which even then was inadequately protected against rodents and predatory fowls.[1] The principal items in his diet consisted of corn, peas, beans, sweet potatoes, pork, and beef, the production of which was easily integrated into the routine of cotton culture. As a result of these circumstances residents of the larger towns were dependent upon other sources than the local farmers for a variety of food items, some of which they produced in home gardens.[2]

The principal winter vegetable stocked by merchants in the towns and the interior villages was the Irish potato. This food was selling at Augusta in May 1848 at four dollars a bushel, and most of the winter supply came from New England. The crop was thought to be adapted only to cold climates and it did not become a common crop in Georgia until after 1850. Potatoes could be grown locally, but the local ones had such a tendency to rot that they were unmarketable.[3] However, an effective method of potato culture had been used in Georgia as early as 1835, when James Camak succeeded in producing three hundred bushels to the acre. He planted them in trenches twelve to eighteen inches deep upon a layer of stable manure. The trenches were then filled with straw and leaves, and covered lightly with earth. When the plants were a few inches high he mulched the ground with a covering of leaves.[4]

Robert Battey in 1849 experimented with roots of all Irish potato varieties common to the vicinity of Rome, and including some from the North. He placed the harvested potatoes in storage and the following spring he found that a few had survived the winter. By planting these tubers and repeating the process of selection for several seasons, he finally obtained a potato with satisfactory keeping qualities. "I am now in possession of a variety of Southern Irish potatoe that is peculiarly adapted to our climate, and by means of which we may in a few years, supply to a considerable extent, the Northern cities, instead of buying from them," he announced. The seed of the Battey potato was sent to leading experimenters in many parts of Georgia and South Carolina. Large quantities of local potatoes soon appeared on the markets and at the agricultural fairs. The agricultural census for 1860 shows a significant increase of this item in the productions of Georgia farmers.

Southern planters had been equally slow to acclimatize other vegetables to local conditions. Like potatoes, onions grew to maturity but did not survive the summer. Onions from Weathersfield, Connecticut, were said to make up two-thirds of all onions consumed in the South before 1850. In 1852, however, George A. Peabody announced, somewhat optimistically, that the Weathersfield had at last lost its monopoly and that onions were then growing in the vicinity of Columbus every month of the year.[5] Likewise the hard cabbage, which once had been considered impossible to grow in the South, was being shipped to the North in the late spring of 1853, along with peas, okra, tomatoes, eggplants, early corn, and melons. The firm-head cabbage, whose development was attributed to the presence of German gardeners in the state, had become almost as common as the long blue collard. Similar progress was reported with turnips and other vegetables.[6]

Robert Nelson played a significant role in developing vegetable crops that would grow, survive, and keep under local conditions. In his nursery near Macon he began the dissemination of such vegetables as rape and colza which he had known in Europe, and the shallot onion which, though brought to America in the early Colonial period, was scarcely known in the South before he brought it into use. He continued this work later at Augusta and Montgomery.[7] He encouraged sending early vegetables to St. Louis and Chicago, to which centers Georgia had access by rail transportation after 1850.[8]

One of the best-known exponents of gardening in the South was Charles A. Peabody. Born at Woodbury, Connecticut, in 1810, he was the uncle of the famous educational philanthropist of the post-war period. A tailor by trade, his formal education was limited, but he pursued through reading an insatiable thirst for beauty and culture. In 1834, shortly after his marriage, he moved to Georgia and settled in the rapidly growing town of Columbus. He put all of his savings into a tailoring establishment, but the transient nature of his clientele, poor collections, and a disastrous fire put an end to this business. He moved six miles from Columbus, to Russell County, Alabama, secured a tract of land there, and erected a four-room log house. Indulging a great passion for agriculture, he began work with his own hands, reared nine children, and became the master of several slaves and the owner of a large house which he christened "Spring Hill."[9] In 1848 he became the agricultural editor of the *Muscogee Democrat* and later he was the horticultural editor of the *Soil of the South*. When that journal combined with the *American Cotton Planter*, he remained the horticultural editor until succeeded by Robert Nelson in 1857. Because of Peabody's influence, these journals put great emphasis upon home gardens, flowers, trees, and shrubs. For every home owner Peabody advocated setting aside a half-acre lot on which to grow the things needed for luxury, comfort, and full living. "When we see every householder in the land sitting under his own vine and fig tree, we shall feel that we have accomplished our mission," he stated to the readers of his horticultural column.[10]

Peabody's best known work perhaps was developing strawberries suitable to the Southern climate. He experimented in 1841 with Hovey's seedlings, acquired from a New York nurseryman. Since the blooms were all pistillate, they bore no fruit. He finally obtained a small yield by planting among the seedling plants alternate rows of wild strawberries whose flowers contained both stamins and pistils. His discovery in this experiment was the basis of his subsequent success in the culture of strawberries.

Observing that the strawberry, once having expanded fully its petals without impregnation, was never productive, Peabody came to the conclusion that the first blossoms were the impregnators. Consequently he cultivated an impregnator which came into bloom early and continued as long as there were pistillate flowers to be fertilized. For the combination he selected Hovey's seedling and the early scarlet, both blooming early in the spring and continuing until late in the summer. From these selections he achieved a production of two hundred bushels of fruit to the acre.[11] Peabody's strawberry was peculiarly adapted to Southern conditions. Known variously as the Peabody, the new seedling Hautbois, and as the Peabody Hautbois, it had been shipped by 1860 to many important markets in various parts of the United States where it was in great demand because of its early arrival and its excellent keeping qualities.[12]

While the practice of alternate planting had been followed by some strawberry growers in the Cincinnati area for two decades, apparently little was known of the actual botanical structure of the American strawberry plant. Nicholas Longworth suspected asexual peculiarities but he did not immediately investigate them. About 1844 he asked the Cincinnati Horticultural Society to conduct experiments on pollenization, the results of which were made public in a famous report two years later.[13] Meanwhile leading Eastern strawberry growers, including Hovey, denied any defect in the male organization of the plants. While Peabody claimed that his discovery occurred eight years prior to that announced by the Cincinnati Horticultural Society, he failed to publish the results.[14] He realized substantial profits from the sale of his plants and he guarded from Northern growers and others the secret of his success.

In 1857 Peabody sent sample strawberry shipments to the editor of *The Horticulturist* in Philadelphia, who reported their arrival with flavor and quality unimpaired, even though they had traveled by wagon, rail, and steamer more than eleven hundred miles.[15] Announcement of the successful shipment appeared to challenge the long-standing monopoly of the Ohio growers. They showed unusual interest in Peabody's durable fruit, but he refused to send a single plant north of the Ohio River for less than five thousand dollars, an offer which Cincinnati growers had no idea of meeting. "Can any strawberry be of value that can be carried twelve hundred miles by a wagon, railroad, and steamboat without mashing?" they parried. It was discovered later, however, that Peabody's strawberry did not thrive well in the more northern latitudes. For many decades it remained the most important parent of all improved strawberries in the Lower South.[16]

In the last decade before the Civil War many truck gardens appeared in Georgia to supply both local and outside markets, the enterprises being largely in the vicinity of towns and cities.[17] Typical was the twelve-acre garden of George A. B. Walker near Augusta. In June 1854 he was growing such vegetables as asparagus, broccoli, cauliflower, cucumber, carrots, parsnips, squashes, beets, leek, celery, tomatoes, eggplants, peppers, cantaloupes, mangoes, rhubarb, and raspberries. He also grew large quantities of early corn, peas, okra, onions, and sweet potatoes, and his orchards produced pears, apples, apricots, peaches, plums, quinces, figs, and grapes.[18]

Important aspects of the horticultural revival of the 1850's was the culture of flowers, the care of shrubs and trees, and the improved construction of houses and farm buildings. Like many non-natives, both Peabody and Nelson keenly appreciated the aesthetic possibilities of the South's flora, and they were critical of the Southerner's penchant for strange and unusual plants.[19] "Our mountains abound in rhododendrons, azaleas, laurels, and thousands of other trees and plants, which only need to be seen to be admired," said Peabody. "Our valleys teem with the magnolia, the bay, the tulip tree, and hundreds of others . . . ." He tried to stimulate interest in an annual flower festival to draw attention to Southern floriculture.[20]

The plantation house of the Georgia Piedmont was embellished with no great variety of trees and shrubs before 1840. A few unattractive specimens of arbor vitae, clinging vines, cape jasmine, and rosemary formed the basis of the landscape arrangement. There was an occasional mock orange, and even more rarely a magnolia. The magnolia grove of the plantation mansion, as it appeared later, was either absent or newly-planted, and if present was inconspicuous because of its immature growth.[21] The flower garden included mainly roses interspersed with dahlias and a few nondescript annuals. Even the rudest cabin was likely to have around it a luxuriant growth of summer roses.[22]

Roses abounded in the greatest variety, being by far the most popular flowering plant cultivated. A University of Georgia professor recalled that in 1810 there were but three or four varieties of roses to be found in Athens, among them being "the *Blush Indica,* the *Damask,* and the medicinal roses." In 1814 an Athenian returning from a tour of military duty near Charleston introduced the *multiflora*. After 1820, however, new varieties came in abundance and by 1840 "the list had swelled into books."[23] On many farms roses were planted as hedges and were sometimes used along board fences to furnish an additional barrier to livestock. The single white Macartney was widely used as an evergreen, while the Cherokee rose was always popular as a fencing hedge. Because fence rails often were in short supply, the hawthorne, crab apple, and honey locust were recommended as substitutes for wooden fences.[24]

Rose and flower enthusiasts were found in all the larger towns and cities. Simri Rose of Macon was a noteworthy example. Born in Connecticut in 1799, he settled at Fort Hawkins on the Georgia frontier where the city of Macon later appeared. In 1823 he helped to establish at Macon the *Georgia Messenger*, and continued with that newspaper until his death in 1869. In 1840 he submitted plans for the landscaping of the fifty-acre cemetery at Macon, which he later filled with trees, shrubs, and wild flowers.[25] His description of this garden in 1849 bears witness to his skill in developing an attractive landscape. He wrote:

> From the river, deep and narrow dells penetrate the ground from fifty to two hundred yards; one of them divides it entirely near its center, through which a rivulet murmurs over a steep and rocky bed to the river . . . . Above it tower giant poplars and the shady beech, and the sun can scarcely penetrate a beam to enlighten this quiet and solemn solitude . . . . Two rustic bridges of rock and earth cross this valley; and in it a pond of about fifty yards in length, by twenty in breadth, has been excavated, supplied by pure water from the springs, and its banks neatly sodded with grass. Around it are several cypresses and weeping willows . . . . A variety of fine roses is also near it, and in perpetual bloom. These are also scattered over the ground, and along the walks and roads in great profusion.[26]

The earliest center of floriculture in upper Georgia was Augusta. Its florists and nurserymen were well known and they enjoyed a wide patronage. Many of these men were foreign born, among them being Prosper Jules Alphonse Berckmans, who, with his father, Louis E. Berckmans, established the outstanding horticultural establishment in the South, known as "Fruitland Nursery." Prosper Berckmans was born in 1830 at Aerschot, Belgium, of a Flemish land-owning family of the lesser nobility. At an early age he was graduated from the University of Tours. The elder Berckmans was a well-known Belgian horticulturist who trained his son in the family tradition. The revolutionary ferment of 1848 caused the family to remove to America, settling first in New Jersey. Recognizing the growing horticultural possibilities of the South, they moved to Augusta in 1857.[27]

In 1860 Berckmans imported an Amur privet hedge from which it is said all the privet hedges in the South have come. His catalogue in 1861 stated that his testing grounds contained, in addition to numerous varieties of fruits, over one hundred each of azaleas and camellias. He was advertising a large number of ornamental trees, flowering shrubs, evergreens, and climbers.[28] Among the shrubs which Berckmans introduced are the golden *arborvitae*, dwarf golden *arborvitae*, and *thuja orientalis aurea nina*. Berckmans supervised the landscaping of many famous antebellum Georgia grounds, such as Rosemary at Newnan, Barnsley Gardens at Kingston, and Esquiline Hill near Columbus. He was the originator and disseminator of more worthy ornamental forms perhaps than any other Southern horticulturist.[29]

The improvement of nursery stock and the growing enthusiasm for floriculture were accompanied by new and imaginative interests in landscape and architectural design. Charles A. Peabody displayed considerable originality in many such developments. He was a devoted reader of architectural literature, and he possessed an intimate knowledge of Andrew J. Downing's works on landscape gardening and cottage residences. The style of landscaping which he advocated for Southerners, however, was original and indigenous. In contrast to Downing's system, it was informal and natural, and sought to utilize native plantings to the fullest extent. It also differed greatly from Berckman's style, with its formal marks of the European art, which had come to be adopted largely throughout the eastern and northern United States. In shunning any exotic system of artistry, Peabody added a unique cultural tone to the sentiment for Southern nationalism so much in evidence during the 1850 decade.

A significant feature of his pattern of landscaping was his insistence on completely reversing the order of emphasis which Downing and other authorities advocated in the landscape design. Downing's system might be characterized as an attempt to harmonize the building with trees, shrubs, and other plantings in its immediate vicinity, the whole to form a somewhat isolated picture to be enjoyed by the viewer from the street or highway. Peabody, on the other hand, would have the building harmonize with the distant landscape, with the house standing relatively isolated in the center of a distant and somewhat circular arrangement of trees and shrubs. "The bulbs and little annuals [should be planted] close under the windows," he advised, "roses and dwarf shrubbery next, then trees." Peabody believed that the dwelling and not the highway should be the vantage point from which to view the landscape picture. In this manner an ostentatious display of the house—a characteristic of an urban industrial society— would be avoided. Thus Peabody was striving to achieve a rural-agrarian expedient for the South, characterized by seclusion within a natural rural environment.[30]

In urging Southerners to capture the natural beauty of the native countryside, Peabody cautioned against cutting down trees which required a century to grow, and he ridiculed the use of foreign shrubs.[31] "We know of a homestead not a thousand miles from here, built by a man of wealth and reputed taste," he said. "The site selected was on the banks of a stream in a setting of large trees. He cut every tree and placed a mammoth mass of timber called a house on top of the hill to bask in the morning, noonday and evening sunbeams . . . as the venerable trees began to disappear, some ornamental bushes were planted . . . . Europe, Asia, and Africa had been ransacked for diminutive shrubs, to take the place of those forest patriarchs . . . . We consider it one of the greatest misfortunes that can befall a man, to possess great wealth without any taste."[32]

Where the grounds were very spacious Peabody would line the avenues with oak trees. "Groups of sweet gum, cedars and wild olives, with a single magnolia, or even a pine, for the eye to rest upon, some distance from the house, will ever be agreeable objects," he said. "Plant only such trees as are long lived, and though we may not live to see the perfection of their beauty, our children may." Southerners believed that the presence of shade trees and thick foliage near the house caused sickness and fever; they did not know about malaria-carrying mosquitoes. But Peabody's advice to plant no tall tree near the house was from a consideration of taste rather than of health.[33]

Other Southern writers on the subject appeared to agree in general with Peabody's theory. Among these was George Kidd of Columbus, who insisted, however, that tall trees should be left near a building in order to bring out its beauty and proportion and to soften the rawness of new construction. He believed that "to dot over the lawn with trees singly or in groups" would be an aid in harmonizing with the distance. William N. White, on the other hand, sought to emphasize a front lawn as the foreground from which the dwelling would seem to spring. To avoid causing dampness and miasma, large trees were not to be nearer the building than thirty feet. White suggested that trees should flank the sides of the building and be joined by a mass of trees in the background. This way of planting would conceal from view unattractive buildings such as the kitchen and servants' cabins. He suggested that shrubs and flowers be selected with a view of staggering the blooming season throughout the year.[34]

Trees were regarded in the frontier tradition as an obstacle to progress and a barrier to agricultural development. The westward movement of the expanding plantation system prolonged this attitude in the Cotton Belt, where it was apparent as late as 1860. Travelers and other observers pointed out the glaring absence of trees in the landscape of Southern towns and much of the countryside. An Englishman described Augusta in 1833 as "a long straggling town . . . with a main street at least a mile long and full of small stores and low taverns." In noting the comparative scarcity of trees, he said that all Southern towns were very much alike.[35] An inn on a Georgia roadside was described as "handsome and costly for a country residence," tastefully and expensively furnished, betraying "an odious aping of town life," but not a tree, shrub, or flower was within sight.

Solon Robinson in 1851 noted an unfortunate lack of public spirit toward beautifying towns everywhere in the South, although at Athens he saw many private mansions resplendent with beauty and adornment. A Texan in 1859 thought the effect of many a handsome village residence in Georgia was lost because a red clay bank too often interposed between it and the street.[36] "It must have struck . . . [everyone] who has ever been abroad at the North, how very bare, unromantic, [and] parched up the generality of our back country villages or country houses appear for want of shade around them," wrote a Georgian in 1857. He described these villages as having broad, sandy streets, and no trees except a few deformed pride of India (Chinaberry) and locusts. He suggested that a variety of trees, including oak, hickory, tulip, elm, and chestnut, be substituted. "I would add for avenues on the plantation or country seat, the Pecan nut, a native of Louisiana and Mississippi . . . the nut of which is very valuable and among the most delicious," he said.[37] He regarded the magnolia as the most magnificent tree known, and one which would richly repay time and labor bestowed upon it.

The average rural residence suggested anything but scented magnolias, and only rarely did it present a spacious, shaded portico to admiring passers-by. The editor of the *Soil of the South*, despairing of efforts to improve the aesthetic tastes of Georgians, described the planter's house as surrounded by Spanish mulberry or Chinaberry trees, "under the shade of which is seen in the summer time a lazy pack of egg-sucking hounds, or noisy sheep-killing curs, half starved."[38] Other documents indicate that the Georgia farm residence was an even match for the low state of its landscape arrangement. "A Northerner who is accustomed to judge of a farmer's property, by his building, would suppose, when he first went into the country at the South, that many of great wealth were poor men, their buildings are so miserable," said Emily Burke, the New England school teacher who lived in the interior of eastern Georgia in the 1840's. She described visits to plantations "where the master's residence had not a pane of glass in the windows, nor a door between the apartments." Nor was there "the shadow of a board to intervene between the ground floor and the coarse, unhewn shingles, as seen on the inside of the roof." Yet the planter's table was loaded with a variety of delicacies and his beds were of "the softest down." His poultry yards were well stocked and the surrounding woods were filled with herds of cattle, horses, mules, and goats, while dozens of red and yellow swine turned up the turf in search of worms and roots. "Yet with all these possessions . . . all [could] be taken down and removed in a few hours," she said.[39] Such observations were not wholly confined to Yankee school teachers. A native of Middle Georgia described farmers of that region as possessing nomadic tendencies. "They seem to esteem a house and its fixtures like an Arab does his tent," said he.[40]

William H. Chambers, co-editor of the *Soil of the South,* warned his readers in the Southwest that those who had converted Middle Georgia from a garden into a desert were destined to make barren the productive fields of those western regions to which they had migrated. "A log house half decayed with age, or a framed house without paint, and . . . a yard . . . without a shrub or a flower . . . are too frequently the insignia of the planter's premises," he said. "Travelling through the country of Virginia we find a different state of things," he continued. "There we see venerable old mansions, comfortable farm houses, and the evidences of wealth all around .... We must fix population before we can educate improve and refine."[41]

Charles W. Howard believed that care and beauty displayed in the homestead would make the lands more valuable and thus insure permanence to the population. "We abandon, without regret, the ill-shaped crazy and comfortless cabin, around which the bare earth burns under the firey [sic] sun, or rank weeds pollute the air with poisonous odors," he said. "But it is a very difficult thing to contemplate the abandonment of a comfortable home ...."[42] Charles A. Peabody thought that the isolation of the planter's house and his limited opportunity for ostentatious display discouraged architectural standards. "Could the dwelling, yard and garden be carried to the market town—be shown at the political gathering, or be exhibited with the fine carriage and clothes, at church, the rural homes of the country would present a different appearance," said he.[43]

Three somewhat conflicting architectural trends are discernible in Georgia between 1820 and 1860. During this period classical architecture, featuring the Greek Revival, was the prevailing standard for men of aesthetic tastes. This style was introduced around 1820 and it was never completely supplanted at any time before the Civil War. However, the appearance of Andrew J. Downing's *Cottage Residences* in 1842 reflected a new trend in architecture in the South as well as throughout the United States.

Downing introduced an element of variety by substituting Gothic picturesqueness for the unvarying formality of the classical forms. He objected to "the false taste lately so prevalent among us, in building our country homes in the form of Greek temples," sacrificing the convenience and comfort of low and shady verandas to the bald display of a portico of stately columns. He advocated the English rural Gothic style with a high pitched roof, arched openings, and side walls of vertical boards and battens. He included designs of the old English, Italian, and Tuscan styles of buildings. His designs of "porches, verandas, or piazzas" illustrated a variety of form and decoration, ranging from the embattled and buttressed portal of the Gothic castle, to the latticed arbor porch of the cottage. "Thirty years ago steep roofs were pulled off and flat roofs put in their places," wrote Daniel Lee in 1848. "Now flat roofs are unpopular and are giving place to steep ones." Lee's characterization was more applicable to the North,[44] whence he had just arrived, than to his adopted section. There were few adherents in the rural South to the Downing school, although some evidence of the Gothic style is found in the contemporary architectural remains of Southern towns and cities. By far the most impressive of these is the old Statehouse at Milledgeville.

A third architectural trend was toward a plantation style somewhat indigenous to the South which did not come into notice until the 1850's. It was a compromise between the classical and the Gothic styles on one hand and the double log house with its simple variations on the other. It emphasized a functional character, simplicity, and comfort. It reflected not only the climate and the agrarian culture of the region, but Southern nationalism as well.

Daniel Lee opposed the Gothic style and believed that classical style houses should be retained with pride and by no means remodeled. He criticized those who squandered their fortunes "on mis-shapen palaces and villas—poor imitations of European aristocracy." The fact that he filled the pages of his journal with illustrations of rural Gothic cottages indicates nothing more than a wide circulation of illustrative plates originating in Northern engraver's shops. A reader of the *Southern Cultivator* called these Gothic house plans "Newspaper Cottages," and criticized them as resembling Yankee farm houses.[45] "Let us have a civilization of our own," said John Forsyth in discussing agricultural buildings, "and depend on our Yankee neighbors neither for refinement and elegance, nor Weathersfield onions and Connecticut cheese."[46]

The editor of the *Southern Cultivator* thought that the Greek Revival might be modified to meet the requirements of the Southern climate. He believed it to be less extravagant than the Gothic but not altogether suitable to plantation use, since it signified the ambition of its owner to live in a Grecian temple. "In costly public edifices, columns are appropriate . . . . But nothing is more like an eagle's feather stuck in the matted hair of a savage, than the frail plank pillars or columns, painted white, so ostentatiously stuck out in front . . . of a dwelling house," he said. One of Lee's correspondents desired neither the "massy Doric columns and Corinthian capitals" for private residences nor the low-roofed porches on a two-story residence, such as Lee suggested. "A three bushel sack of bran called a bustle, did not more disfigure a pretty woman, than does [sic] the lofty fluted columns, erected at vast expense, and in violation of all propriety and utility," he said. "This style of building and that of two story houses with sheds to sun beds on, painted white . . . need reformation altogether."[47]

Thus the Greek Revival style and the newer Gothic forms in

the early fifties began to merge into a new concept of a farm house. This native development harmonized many of the conflicting elements between the two styles. "The house should be, not an imitation of some showy village mansion, with porticoes and Ionic columns, or of some cocked hat cottage, all gables and no house, but should be moderate in proportion . . . with no expense for mere ornament," wrote a Floridian in 1852.[48] A South Carolinian also rejected both extremes, believing that emphasis should be given to a practical house for people of moderate circumstances. He characterized the Gothic as "a feudal hybrid of a stunted church." He also thought that Grecian architecture, for which a great partiality existed, was "too stiff to associate with trees," and its heavy pillars, unbroken shadows, and "heavy frowning entablature" were uncongenial to the rural countryside.[49]

An editorial on rural homes by Charles A. Peabody gave scant attention to what he termed "gentlemen's country residences," but went much into detail concerning the homes of the "bone and sinew of the State." Unlike many of his contemporaries who sought to modify the Grecian and Gothic forms, Peabody attempted to inspire improvement in the rugged houses of the average farmer. "There are many heavy, double log cabins . . . that have absolutely cost more money than would a light and beautiful cottage, ornamented with graceful colonnades, airy porches, and latticed windows," he said. He advocated, "not a gew gaw palace, or a gingerbread cottage, but a substantial, comfortable, home," adapted to the means of the owner and to the climate of the section. "Give it that graceful colonnade and airy porch for our hot summers, and yet the close and comfortable room for the chilling, changing season," he advised.[50]

In keeping with this idea a correspondent in the *Southern Cultivator* believed that emphasis should be placed upon the kitchen instead of the parlor in constructing farm houses. "We do not mean the kitchen in the great house," he explained, "where lazy servants have entire control, and the lady of the house never sets her foot within its precincts, but the homely, comfortable kitchen of the well to do working man, where the wife and the tea kettle sing together."[51]

The architectural enthusiasts in this age of Southern nationalism demanded the full use of native materials in the construction of houses. They declared that it was just as important to use native lime, granite, slate, brick, and cement as the timber from surrounding woodlands. Builders in Floyd and adjoining counties made use of local slate for roofing, and the kilns of Cass County for lime and plaster. A few structures of granite appeared in northeastern Georgia. Field stone and local hand-made brick were used in chimneys as well as in foundations, while native oak and long-leaf pine went into the main structure. Wide boards whose surface had been planed by hand and painted, sometimes to simulate marble, were to be found on the interior. Plaster and wall paper were used to a lesser degree.[52]

The decade of the fifties was the most prosperous period that Southern agriculture experienced prior to 1860. It was a decade in which many substantial log houses were transformed into larger, white-columned structures, with the original logs left underneath to betray the transformation to later generations. Such is the history of Charles A. Peabody's "Spring Hill" near Columbus, and Charles W. Howard's "Spring Place" at Kingston.[53] Hundreds of other structures throughout western Georgia show a few trappings of the Greek Revival, superimposed upon a framework of logs. Quite frequently in the center these have a great hall or breezeway which once joined the double rooms of a log cabin. Sometimes the simple first-floor structure of a log cabin was duplicated in a second story, while two full columns, flanking the entrance to the hallway, formed the only feature suggestive of the Greeks. The exposed end-chimney was a traditional feature of these houses, a carry-over from the original style of the log cabin. A lean-to back room of a single story might have been added in time to accommodate a growing family. Few suggestions of the Gothic influence were to be seen in the newer areas of the Piedmont Cotton Belt.

Architectural survivals as well as numerous documents furnish ample evidence that, in 1860, all wealthy planters by no means lived in great mansions with stately columns, as pictured by later romantic writers. A visitor to Georgia at that time was surprised to find one of the wealthiest and best-known planters in Middle Georgia neatly dressed in a summer suit "sitting on the very unpretending porch" of his home. The residence of Farish Carter in Baldwin County likewise failed to bespeak the great wealth of its owner. An early description of "Liberty Hall" indicates that the ante-bellum home of Alexander H. Stephens adhered with remarkable fidelity to the system of natural landscaping and practical architecture.[54]

Interest in an improved style of the country residence inevitably led to a consideration of barns and other plantation buildings. When Frederick Law Olmsted observed that there was hardly a poor woman's cow on Cape Cod that was not better housed than a majority of the white people of Georgia, he might have added that most cows in Georgia were not housed at all.[55] "Not one planter in a hundred has a house which can be dignified with the name of a barn," wrote the editor of *Soil of the South* in 1854. He described the usual arrangement as consisting of a rail-pen for shucks and fodder, and oats left stacked in the fields. "A Massachusetts farmer wouldn't ask a better living than the loss about a Georgia farmer's barnyard," he said.[56]

A Georgian in Lee's army, writing from Chambersburg during the Pennsylvania campaign of 1863, showed both surprise and admiration for the fine appearance of the farms and homesteads of that region, indicating that order and neatness in a rural setting were entirely new to him. He greatly admired the large and substantial barns of the German settlers, and the fine state of cultivation of their fields, but he believed them to be "very far behind the age in intelligence."[57] Peabody thought there was nothing in which Georgia's agricultural reputation suffered so much as in farm buildings. "From the dwelling house down to the wagon shelter, all of our improvements are ordinarily of a temporary nature," he said.[58]

Some improvements in farm buildings were noted during the 1850's. Robert Battey built a model barn at Riverback, and exhibited the plan at the Macon fair in 1851. Labeling it the best barn in North Georgia, Dennis Redmond stated that he had not seen more than half a dozen barns worthy of the name in all of that region. The *Southern Cultivator* carried an engraving and a full description of a model barn built by John Bonner of Greene County in 1852. The main structure was eighty-four by forty-four feet, to which was added a wing thirty-two feet square wherein was installed a horse-powered engine for the feed mills. There were twenty stalls on the ground floor, above which were two stories. The middle floor contained ample storage for hay, straw, and fodder and space for corn shellers, straw cutters, and thrashers. On the top floor was the grist mill. The prepared feed was dropped through a scuttle hole into a box wagon on the ground floor and conveyed along a passage to the stalls. "On rainy and wet days, I can thrash my grain, and cut, grind, and prepare food for my stock; thereby making those days as useful as fair ones," said the proud owner of what was perhaps the most efficient barn in Middle Georgia.[59]

Planters also evinced interest in their slave quarters. A physician

GARDENS AND BUILDINGS

at Augusta condemned the practice of building Negro houses too near the ground, and with faulty chimneys which filled the rooms with smoke. The editor of *Soil of the South* recommended single cabins sixteen by eighteen feet, constructed of logs with the openings between them covered with boards instead of daubed clay. He suggested that the house be raised at least two feet above the ground in order to give free circulation of air, to prevent dampness and to facilitate periodical cleaning of trash and filth from underneath. Brick chimneys and shingle roofs were recommended, with at least one glass window and two ventilation openings above the door. To minimize fire hazards cabins were to stand in rows one hundred feet apart with fifty to seventy-five feet between structures.[60]

Improvement in smokehouse construction was thought to be one of the greatest needs of the average planter. "A filthy smokehouse is a disgusting subject to write about, but as they are so numerous, I hope to be pardoned," said one. "It is enough to restrain the most inordinate appetite [to] be shown into the smokehouse, and regaled with the scent from its ground floor, spread with fragments of flesh and bones, and decorated with fat cans and soap gourds; my word for it, smoking ham and dainty steak would have no attractions for them."[61]

This planter's ideal smokehouse was a square building set upon pillars four feet high, with a height of nine feet from the floor to the ceiling. Two feet below the ceiling were two small openings for the admission of air. Smoke was admitted to the center of the building through a stone pipe passing through a water container to condense the steam, thus keeping the fumes cool and dry. "The object of smoking meat," he reminded, "is to drive off flies, dry up its moisture ... and to deposit on it a quantity of pyroligneous acid as a preventative of insects." He advised the use of perfectly dry wood, preferably hickory, so as to produce the maximum of smoke with the least heat. Dirt floors were discouraged since meat drippings produced stench and dampness. A wooden floor covered with three inches of sawdust, over which lime could be sprinkled, was recommended.[62]

In the spirit of Southern nationalism, many wealthy planters boycotted Northern watering places and resorts. There followed an effort to make the plantation a more comfortable and pleasant place for living during the hot summers. Some planters whose houses were located near a spring of water, with sufficient elevation and adequate volume, installed hydraulic rams. With kaolin pipes made at Augusta, they succeeded in bringing running water into the houses, and occasionally to a fountain spray on the lawn. Ingenious devices were used as protection against flies and mosquitoes, constant pests during summer months. Huge fans swept back and forth by Negro servants, and screens made of cotton netting, kept insects from the table and helped to make mealtime a more agreeable occasion. The location of the kitchen in a separate building was also an advantage; it lessened fire hazards and removed the disagreeable odors that sometimes emanated from this servant-run establishment. Flies were generally tolerated in ante-bellum days, however, except by the most discriminating. The mosquito problem was mitigated by placing bedrooms on the upper floor and by using cotton netting. The inhabitants of the humbler homes resorted to the burning of rags. Sometimes a few slices of fresh beef were placed beside the bed in an effort to lure away those bloodthirsty insects.[63]

It was possible for some planters to enjoy ice on their plantations during the summer. Ice cut and stored during New England winters was shipped to Southern ports and thence by rail to interior centers. In a well-constructed icehouse, it could be kept on the plantation the entire summer. Ice retailed at Augusta during the late summer of 1855 at two cents a pound. It could be delivered at Macon, Columbus, Augusta, or Atlanta early in the summer at less than half of this cost.[64] Some planters joined with neighbors to purchase a supply at wholesale prices.

Those who sought the luxury of summer ice built an ice house on porous soil preferably on the north side of a hill, where a six-foot-square hole was dug ten feet deep. A wall made of four-by-four studding, boarded up on both sides, was built into the excavation. The four-inch space between the encasing walls was filled with spent tan bark or sawdust to provide insulation. The floor was constructed of boards covered with sawdust. The roof was covered first with straw to a thickness of two feet, and over this was a covering of shingles. If the soil failed to absorb the water from the melted ice, a small drain pipe was laid to an opening on the hillside. Such a house filled with ice in the later winter was said to supply an ordinary farm for a full season.[65]

Houses, together with their setting and surroundings, were intimately a part of the whole structure of the lives of people who inhabited them. They reflect the social and cultural aspects of the period in which they lived. It was the enthusiastic attention given to buildings and gardens, largely in the last decade before

1860 which gave the Cotton Belt in Georgia its greatest claim to a settled and serene life in a rural environment. However, this ideal often has been emphasized by later writers of romantic bias somewhat out of proportion to historical realities.

XII

Cotton, Corn, and Slavery

THE COTTON PLANT was wholly new to the improved system of crop rotation which had evolved in Europe by the time the Cotton Belt developed. As a result, the Southern planter experienced some difficulty in integrating his principal staple to a system which stressed diversification. About 1835 Thomas Spalding adopted at his plantation on Sapelo Island a system which he claimed was the nearest approach to Flemish culture to be found in Georgia. This consisted merely of planting a row of corn and a row of cotton alternately on permanent ridges and keeping the entire field free of other growth.[1] In the Piedmont cotton was usually planted in drills four feet apart, the distance between the plants ranging from eighteen inches to four feet, depending upon the fertility of the soil.[2]

The fact that both cotton and corn were clean culture crops not only limited their use in a system of rotation with grasses, but also encouraged erosion and left the soil deficient in organic matter, particularly on rolling land.[3] Largely because of Edmund Ruffin's influence, the earliest efforts to restore fertility to cotton lands consisted of adding limestone, ashes, and other alkalies, but it was soon discovered that cotton would thrive on unlimed soils.[4] By 1860 it was generally understood that lime provided only one of several elements necessary for restoring fertility to cotton lands. Many farmers recognized the beneficial effects of leguminous plants and practiced what they called green manuring, although the nitrogen-fixing qualities of the various legumes were not understood. Despite increased yields from plowing under certain legumes, planters were reluctant to forego a cotton crop necessary for sustained soil-building programs.[5]

The most widespread soil amendment used before 1850 was derived from the compost heap. Yet the use of this additive was far from universal, since the clearing of fresh woodland was the usual solution to the problem of wasted land. A farmer in Middle Georgia showed his aversion to manuring land when he said "that it was too small a business for him to go about dropping a shovel of manure here and there" and that he had rather be seen with an axe on his shoulder than perched on top of a manure pile.

Into the compost heap went all manner of offal, straw, leaves, ashes, swamp muck, soap suds, as well as manure from the stables and barnyard. Frequently the compost heap was constructed to permit seepage to drain into a cistern where liquid manure was collected. Livestock were corralled on barren areas to aid in the rehabilitation of these soils, although this method was not used on the stiff clay soils of the Piedmont where treading caused packing and clod-forming in wet seasons.[6] Animal manure alone was generally too scarce for extensive use, but after 1850 planters came to recognize the value of returning the cotton seed to the soil as a cheap fertilizer.[7]

The increased emphasis on soil improvement after 1840 caused considerable discussion concerning the practicability of using "night soil," a fertilizer made of human excrement. This substance received commendation in the movement to achieve economic independence, particularly after the first importations of Peruvian guano. One planter produced figures to show that an adult slave would produce annually over five hundred pounds of "domestic African guano" the value of which he claimed was more than double that of the food which he consumed. However, night soil failed to produce the same results as guano. Being over-charged with salts, it had to be mixed with imported gypsum, charcoal dust, and sometimes salt, for drying and de-odorizing. To fix the volatile ammonia, quick-lime, unleached ashes, or sulphuric acid were used.[8] Poudrette was an artificial manure made by composting dried night soil with powdered charcoal, swamp mould, and lime. Its exact composition was kept secret by the manufacturers. Dried and ground into a powder, it was commonly sown in drills at the rate of twenty-five bushels to the acre.[9]

These fertilizers were never popular in the Cotton Belt, being limited "by false delicacy and silly squeamishness," according to one advocate of their use. The cumbersome inefficiency of the compost pile, and the absence of grass and livestock for sufficient stable manures, led to an increasing interest in commercial guano during the 1850's. Early shipments of guano came in bags containing many lumps which had to be sifted out and crushed on a plank floor. The sale of highly adulterated guano greatly retarded its early acceptance by farmers, but in time they became fair experts at judging its quality by the proportion of feathers and shells which it contained.[10] Its original cost in 1845 was $60 a ton. By 1850 it could be bought at Savannah for $45, and railroads were delivering it to interior points at such special rates as to reduce the total delivery cost to slightly under $50. While this price was still too great to promote its general use, Piedmont farmers quickly recognized its possibilities.[11] Its high portability enhanced its value in areas where a dearth of barnyard manure existed. "We have been much encouraged since the introduction of Guano, and it is perculiarly suited to us in the South," wrote John Cunningham of Greensboro.[12] However, he suggested the continued use of manure and compost except on the more remote fields.

Advocates of the muck heap died slowly and hard, to be sure. Jarvis Van Buren thought that it was "superlative folly to be spending money for one article whose greatest value [was] being far-fetched and dear bought," when every man could collect manure on his own premises with little cost. He predicted that guano would prove advantageous only in town gardens. Edmund Monroe Pendleton of Sparta believed that it would hasten the economic ruin of the South to be transferring to others the bulk of the profits from slave labor. Others predicted that a continued use of guano would produce a hopeless soil sterility.[13]

Despite dark forebodings the use of commercial fertilizers was well under way in Georgia by 1860. As early as 1853 efforts were made to secure direct importation from South America, but a shortage of orders caused the failure of this enterprise. Six years later, however, during the first four months preceding April 1, 1859, the Central of Georgia Railroad alone hauled over twelve hundred tons of guano to Macon where it was distributed by wagons to farmers in that vicinity.[14] In 1860 it was being advertised widely in journals and newspapers, and its use was enthusiastically advocated by agricultural editors. Among the cotton-growing states Georgia was recognized as a pioneer in the use of this new fertilizer.[15] "It needs no brighter imagination than the one that can consider yon hooped out skirt to contain a live woman, to imagine that at no distant day our farmers will be feeding their plants with small pills, instead of, as now, the cart and team load to a few hills," wrote a Georgian on the eve of the Civil War.[16]

Commercial guano received its greatest impetus in Georgia by its successful application to Hancock County soils as reported by David Dickson and Thomas J. Smith. The former by 1860 had spent $38,000 for the product.[17] To Dickson, more than any other man in the South perhaps, planters looked for guidance in a revolutionized system of cotton culture during the two decades before Appomattox. His contemporaries were practically unanimous in acclaiming him the most successful cotton planter in the state. A fortune of more than half a million dollars, which he accumulated in the fifteen years before 1860, gives some substance to this distinction.

While Dickson had limited educational advantages, he had a great respect for learning and he was an avid reader of agricultural journals. Losing his wife early in life, he remained legally unmarried afterward, although it was generally known that he had a daughter by his mulatto housekeeper. Suffering some degree of ostracism by his neighbors, he claimed to care little for social prestige and devoted himself solely to accumulating wealth. A visitor to his plantation in 1859 found him a genial and hospitable host.[18]

Unlike other men of great wealth, Dickson engaged in no major enterprise other than farming. He began his operations in 1845 with an inherited capital of 25,000 acres valued at nearly a quarter of a million dollars, in addition to slaves, livestock, and investments. The center of his agricultural operations was in the lower half of Hancock County where the land was relatively barren and sandy and required liberal soil amendments to produce good yields. His experiments with commercial fertilizer began in 1846.[19]

Dickson's method was to plant the best fourth of his land in cotton for two to six years, depending on its fertility. He then put it in corn for one year, following with small grain, and then a year of fallow. As a general rule, each fourth of his arable land was in cotton, corn, and small grain, with the remaining fourth lying idle. He removed nothing from the field except harvested crops, and he returned the cotton seed to the soil. He plowed the land on contour rows to an average depth of seven inches, completely turning under all cotton stalks and other litter left on the ground. Observing that virgin land containing considerable mould never baked, and withstood long droughts as well as wet seasons, he concluded that land should be kept in a condition similar to its virgin state.[20] Thus he was able to ameliorate somewhat the problem of soil depletion and wastage through erosion.

While Dickson was the most extensive land owner in his county, he was among the more conservative in his use of slave labor, keeping only the more intelligent and skillful Negroes. Unlike many of his contemporaries, he did not invest his surplus in slave property and was not averse to employing white laborers for farm work. His chief outlays were for fertilizer and land. "Double the productiveness of the land, and it will be worth four times the present value," he wrote. "Double the number of slaves and the price will depreciate one-half." To increase productiveness, he used 175 pounds of guano to each acre of cotton, with smaller amounts on corn and small grain. His per-acre production was increased by shortening the distance between plants. He was able to cultivate 48 acres of cotton to the hand "without driving either Negroes or mules, or disturbing his own quiet and ease." He improved his cotton through selective breeding and sold the seed at $5 a bushel throughout the Southern states. A Tennessean who had planted the Dickson variety for two years pronounced it "the best Cotton by 250 pounds per acre" that he had ever planted.[21]

Something of the scale of Dickson's farming is indicated by the fact that his stock of cattle in 1860 consisted of 330 head, of which 100 were milk cows and 30 were work oxen. He had 600 hogs, 200 sheep, and 47 horses and mules. The manure from these animals was applied to the land, together with $9,000 worth of guano. His Hancock County farm consisted of 13,000 acres in one body, nearly a thousand of which were in cotton and 860 in corn. On this farm he worked 55 Negroes and three seasonally-employed white men. From the labor of less than 60 hands he gathered 667 bales of cotton, averaging 425 pounds, and sold $5,500 worth of corn, wheat, and livestock products. He produced 180 tons of hay, 18,000 bushels of corn, 8,100 bushels of small grain, 500 bushels of peas and beans, and 250 bushels of potatoes in addition to such sundry items as wine, honey, butter, and wool. Almost everything used on his farm, including food, wagons, carts, plows, and shoes, was made on the premises. Despite very heavy expenditures for guano, he claimed that his profits were never less than twenty-five per cent on invested capital.[22]

While David Dickson's farm was considered the model agricultural operation in Middle Georgia, there were other farms in Hancock County which were worthy of notice. One of these belonged to James Thomas, who, with 25 hands, cultivated 350 acres of corn and 380 acres of cotton, in addition to wheat, oats, and vegetables. Thomas M. Turner in 1858 had $34,000 invested in Negroes, land, and equipment. By imitating Dickson's methods, he claimed that year a profit of 16½ per cent on invested capital. His gross income came from cotton, corn, bacon, wheat, potatoes, oats, and forage. On Shoulderbone Creek, in the northern part of the county, were the so-called "brag farms" of Dimos Ponce, John Bonner, Miles G. Harris, and Isaac Whitten.[23]

David W. Lewis described Hancock County in 1859 as a paragon of progress and achievement in the science of farming. He described the community's "country colleges," its white mansions, gardens, and orchards, with all the unmistakable signs of comfort and taste about them. He stated that Devon and Durham cattle had replaced the bony and dappled common herds and that homemade plows had banished those produced in New England foundries, such as the Boston Blue Mason and Ruggles. It was claimed by the Savannah *Journal and Courier* that no county in Georgia could produce more agricultural skill, intelligence, and refinement than Hancock. This sentiment prevailed throughout Georgia and adjoining states, where Dickson's farm was often referred to as the "Modern Mecca."[24]

While Dickson and his neighbors were perfecting the new techniques of cotton production on the poorer lands of the lower Piedmont, significant attempts were being made elsewhere to improve the quality and the market value of the staple. Unimproved gin saws broke the fibers and failed to remove all the lint from the seed, making the cotton less attractive to buyers and the seed unpalatable to livestock.[25] There were few improvements in the cotton gin before the Civil War. With the exception of the gin feeder, which eliminated the work of a single operative, such changes as were made involved the size and shape of the stands and the method of transferring power to the machinery. In 1850 it was still customery to gin cotton on the plantation where it was grown, but the privilege of custom ginning was extended to a few small growers in the neighborhood. Ten years later portable horse-powered gins were common, and stands contained from forty to sixty saws each. The best gins, equipped with feeders, cost approximately $385. The maximum production of one gin stand in a day's operation was seven bales or 3,000 pounds of lint. This required the work of a woman to feed and attend the gin and two boys to drive the mules.[26] One ante-bellum observer described a Georgia gin house as a place of "eternal clamor, clangor, clatter, clang, jumping, jarring, and back lash." Friction on poorly constructed gearing, often made of wood, caused the rollers to burn out rapidly, and the machines were constantly in need of repair.[27]

Up to 1850 round cotton bales were still common in Georgia, although seldom seen elsewhere. This pattern was established by the early plantation practice of packing cotton into bags by the hands and feet. The earliest screw presses were made of iron and were used at Augusta for repacking cotton for shipment by boat. By 1820 wooden screws were made on the plantations and in time they came into general use. These were considered an improvement over the iron screw, which often broke under pressure and sometimes caused serious accidents. There was not much uniformity in either the size or shape of cotton bales until after 1845 when the Savannah Chamber of Commerce urged planters to adopt square bales of 450 to 500 pounds in weight. They were to be packed to a uniform size of five feet and four inches long and twenty inches wide, covered with good bagging and well roped. "A ship from Georgia is easily recognized in Liverpool by the mass of loose and soiled cotton accumulated in the hold," it was stated.[28] Agricultural leaders condemned "false packed" bales which contained discolored and inferior lint packed in the middle of the bale. A Georgian in 1852 suggested that cotton be pressed in close bulk for four weeks before ginning to allow oil from the seed to be imparted to the lint, thus to increase its weight and to give it a cream tinge which enhanced its market value. To prevent oil from being evaporated by wind and sun, cotton picking was begun as soon as one hand could gather fifty pounds a day.[29]

The techniques of marketing the cotton, which was an export crop of considerable magnitude, presented serious problems to growers. The wide variation in the price of the staple after 1820 sometimes wrought disastrous cycles of depression and prosperity. The individual planter was unable to predict prices, since the staple was subject to outside factors over which he had no control, and about which he had little knowledge. These factors included probable production and world consumption, and carry-over, concerning all of which the grower had to reckon with unreliable figures. New and unforeseen factors also disrupted his calculations. In 1853, for example, there was speculation concerning the probable effect of the book *Uncle Tom's Cabin* on the world cotton outlook. This speculation was stimulated by a quotation from the *London News* which suggested indirect aid to the abolitionists by developing cotton production in Egypt, Australia, Africa, and the West Indies.[30]

In 1844, when the price of cotton reached the lowest point that three generations of growers had ever known, there was general recognition of the need to curb production in order to improve the price and to release land and labor for other enterprises. At this time Dimos Ponce, a former Floridian of Spanish lineage, living near Mt. Zion in Hancock County, advocated a scheme of crop control similar to that of the New Deal almost a century later, even including state, county, and individual farmer allotments. The details of his program were first presented in a letter to the Hancock Planters Club of which he was a member, and later published in various journals.

Ponce called for an agricultural convention of the cotton states, with delegates who would have authority to determine some method by which there could be an enforceable system of crop reduction. He suggested the penalty of confiscation for overproduction, and he advocated that unused cotton land be devoted to livestock and subsistence crops.[31] He maintained that withdrawing a third of the usual labor devoted to cotton production would result in a more intensive cultivation of the remaining cotton land. "We shall be able to obtain at least 9 to 10 cents per lb.: netting the farmer one fourth more money than the whole of the present crop supposing it to average him $5\frac{1}{2}$ cents," he wrote. Ponce warned that unless cotton planters agreed to such a plan the price of cotton in the next two seasons would range from two to four and a half cents.[32]

Subsequently steps were taken to organize a state agricultural convention at Milledgeville for the purpose of carrying out Ponce's suggestions and other programs. Few planters responded to the call for a Milledgeville convention, however, and the matter lagged. County meetings were held in a few communities throughout the state, but no unanimity was reported. It was obvious that success of the scheme depended upon similar action in other states. At a meeting in Harris County resolutions were passed whereby planters agreed "to diminish the quantity of cotton by each of them raised, so far as to produce all other things necessary for home consumption and to supply those in the county not engaged in agricultural pursuits." It was claimed that many planters remained absent from the meeting at Hamilton because they feared that they would be asked to give a written and binding agreement to reduce acreage.[33]

Prosperity returned after 1849, and the decade which followed brought a gradual upswing of prices. However, Ponce's idea of crop reduction persisted throughout the 1850's. Commenting in 1851 upon the proposed cotton planters' convention at Macon for the purpose of discussing the stabilization of cotton prices the *Rome Courier* stated:

> These experiments have been tried before, and found quite unsatisfactory. If this whole interest could be confined to the control of 100 cotton planters of honor and honesty, the cause would appear more hopeful; but to suppose that the resolves of a cotton convention, will bind the plows and hoes of one hundred thousand cotton planters, or cause them to move at their bidding is absurd. Suppose every cotton producer in Georgia should agree to raise no more cotton until the price appreciates to 10 cents; how many Mississippi and Texas planters would commend their wisdom or follow their example?[34]

In 1862, Ponce again promoted the idea of crop reduction, this time for the purpose of providing subsistence for the Confederate armies. He reported a meeting at Sparta at which a large number of farmers signed an agreement "to abandon entirely for the present the culture of cotton, and to devote their whole resources to the production of provisions." The chairman was ordered to appoint a committee from each militia district in the county to present the matter more fully to all farmers and to secure their cooperation. The Hancock club then called upon all farmers throughout the Confederacy to follow this policy. Later in the year the state legislature enacted a law designed to reduce cotton production, but like the various sumptuary and price control measures of the war period, the law lacked the machinery for effective enforcement.[35]

While some Georgians were mastering the new techniques for the production and marketing of cotton, others were devoting their attention to corn. The latter was not grown for export, but it was in the eighteenth century the most important crop grown in the state and was throughout the ante-bellum period second to cotton in importance and in the total value of crops produced. Because of the emphasis on cotton in the pre-war years and the expansion of that crop, very little had been done to improve the culture of corn. James Camak of Athens made a significant discovery when he succeeded in producing seventy bushels of corn to an acre by abandoning plowing and using a mulch of straw and leaves to control grass and to hold moisture in the ground. Surmising that the corn plant had a delicate root structure, he concluded that the prevailing practice of deep plowing close to the plant wrought serious injury to its growth. He theorized that corn should be grown with little or no plowing. In a corn field infested with crab grass, with the young blades beginning to turn yellow, he spread a small quantity of stable manure around each plant and then covered this with wet leaves. The grass died and the corn flourished. Without further cultivation it produced forty-two bushels to the acre. Camak published the results of his experiment in the *Farmers' Register* in 1835; however, few Georgians were subscribers to agricultural periodicals at that time, and the method remained generally unknown to them.[36]

Thomas J. Dickson, a brother of David W. Dickson, observed a poor neighbor who had only a hoe with which to complete the cultivation of a crop of corn. After the neighbor made an abundant crop and Dickson's corn crop failed as a result of drought, Dickson came to the same conclusion at which Camak had arrived earlier. He concluded that corn land initially should be plowed deep and pulverized; afterwards it should be cultivated to no greater depth than would result from use of a hoe which merely uprooted the grass and weeds and mulched the surface.[37]

For cultivating corn on this principle, Dickson perfected a sweep twenty-six inches wide with the end of the wing designed to run close to the young corn. The seed was planted in the bottom of very deep furrows seven feet apart, and then lightly covered. On land not well drained the seed was planted on ridges, and in the middle of the furrows was planted a row of peas. Subsequent cultivation consisted of three workings with the sweep. By such cultivation, and with the use of guano, an acre was made to produce twenty to twenty-five bushels of corn. On sandy soils the sweep developed by Dickson was used more extensively. On the stiff clay soils of the upper Piedmont the after-culture consisted of additional deep plowing with scooters while the plants were still small, then a rapid graduation to plowing of succeeding shallowness. The final work was usually done with a hoe.[38]

This system of corn culture, involving deep plowing and heavy fertilizing in the preparatory stages and shallow plowing thereafter, became traditional in Georgia and basic tools for cultivation developed in this period remained unchanged for many decades.

The system was popularized by various agricultural writers, including Charles A. Peabody, James M. Chambers, Garland D. Harmon, and Edmund M. Pendleton.[39] Chambers and Pendleton recognized the advantage of leaving the fodder unharvested on the stalk to improve the quality of the grain, but because of the paucity of forage for livestock, this practice was seldom followed. Harmon's use of the Georgia system in Mississippi and Louisiana and his numerous articles on corn culture helped greatly in popularizing this method in the Southwest.[40]

The successful use of guano and improved cultural methods of cotton and corn, together with improved markets, caused a revival of interest in the institution of slavery during the fifties. The management and treatment of slaves, and the overseer question, took on added importance. This new interest was accompanied by a lively resurgence of the slave market. During the decade prior to the Civil War, slaves rose in price out of proportion to the relative value of cotton. Field hands in 1859 often sold for as much as $1,700, and a good blacksmith brought $2,000, while the average price of cotton during the 1850 decade was between ten and twelve cents.[41] "The old rule of pricing a Negro by the price of cotton . . . that is to say, if cotton is worth 12 cents, a negro man is worth $1200 . . . does not seem to be regarded," wrote the editor of the *Federal Union* in expressing apprehension of disaster at the boom in slave prices. In the vicinity of Griffin planters on exhausted lands were said to be sustaining themselves largely through the increase of their Negroes.[42]

Slave management on upland cotton plantations before 1850 was highly unscientific. Earlier writers on this phase of plantation management in Georgia dealt largely with conditions on the rice and sea island plantations along the coast, where life and labor were hazardous and mortality higher than on the cotton plantations of the Piedmont. The nature of rice culture permitted little variation of the traditional routine. Some attention was given to the internal system of policing slaves, but food, medicine, and work discipline were the principal topics discussed. Roswell King, manager of the Pierce Butler estate, was among those who recognized the importance of keeping the slaves healthy and contented. To accomplish this he permitted some degree of free enterprise and provided food allowances. "I find at Butler's Island, where there are 114 little Negroes, that it costs less than two cents each per week, in giving them a feed of Ocra [*sic*] soup, with Pork, or a little Molasses or Hommony or Small Rice," he wrote. "The great advantage is, that there is not a dirt-eater among them . . . ."[43]

It was a general practice on many plantations in Liberty County to allow old Negroes to administer medication to the young. They also indulged in blood-letting, administered purgatives, and prescribed hot baths and tonics. They also extracted teeth on occasion, sometimes cutting around an aching tooth with the same lancet used in blood-letting. In cases of ordinary fever the practice in vogue included letting "a pint of blood to reduce the pulse, then administering ten grains of calomel, followed in the morning by half a teacup of castor-oil containing three or four drops of turpentine to impart additional potency to the dose, and finally snake-root tea to brace up a halting system . . . ."[44]

The health and general welfare of slaves took on added importance during the 1850's. Many believed that the Negro's susceptibility to such diseases as pneumonia, tuberculosis, typhoid, and malaria was a result of his unique physiology. Garland D. Harmon was among those who insisted that some Southern physician should write a book on Negro diseases for the use of plantation managers. Dr. John Stainback Wilson of Columbus agreed in 1860 to undertake such a task.[45] Well known as the author of *Woman's Home Book of Health* and a health column in *Godey's Lady's Book* as well as numerous articles in agricultural journals, Wilson was somewhat ahead of his contemporaries in medical theory. He believed that the common practices of blood-letting and purging of patients suffering from malaria and typhoid were unwise, and often resulted in the death of the patient. He emphasized the importance of better food, clothing, and housing in effecting improvement in the health of slaves. "There can be but little doubt that the dreaded typhoid fever," he wrote, "owes its origin largely to the accumulated filth of years about negro cabins."[46]

Despite the approximate validity of these theories, Wilson's concept of medical practice in slave quarters was based upon what he called the Negro's defective heat-producing powers. His susceptibility to pulmonary diseases was attributed to his limited breathing capacity. He claimed that the Negro's tendency to sleep with his head to the fire was a natural instinct to avoid taking cold air into his lungs. Wilson insisted that pork and corn were heat-producing foods, and these items were essential in the Negro's diet. But, said he, "while fat meat is the life of the Negro . . . it is a prolific source of disease and death among the whites of the South and West."

A large amount of fat pork in the diet of Negro children was generally thought to prevent infection and death from intestinal worms.⁴⁷ A liberal ration for an adult slave in the Piedmont consisted of three and a half pounds of middling meat, or four pounds of pork shoulder per week, one and a half pecks of meal, with some field peas mixed with a little molasses or sorghum. "The more exhausting the labor the fatter the meat" was a common axiom. The Negro's distaste for beef and mutton was widespread, probably because he had never tasted fat beef or fat mutton. "They rejoice in grease," said Charles W. Howard. "They do not relish lean meat of any kind."⁴⁸

Some planters urged that a week's supply of food be distributed to the Negroes at one time. One who claimed great managerial experience rejected weekly distribution of food because it would encourage waste, stealage by other Negroes, and uneven consumption throughout the week. He also objected to Negroes' doing their own cooking since it required time needed for rest at noon and in the evening, and resulted in poorly cooked and indigestible food. Where there were as many as ten hands to be fed he thought it wise to assign one person to do all the cooking.⁴⁹

A custom that had become general on many large plantations by 1860 was the annual barbecue, usually on the Fourth of July, at the "laying by" season in late summer, or at Christmas time. This festivity was promised as a stimulus to early completion of the work of cultivating the crop, or finishing the harvest. The outlay of food served by John Woolfolk of Columbus to his Negroes in 1853 at their annual barbecue consisted of two each of cows, hogs, lambs, and goats, in addition to hams, chickens, chicken pies, pig head stews, onion sauce, Irish potatoes, beets, squash, green corn, tomatoes, cucumbers, and other vegetables, with watermelons for dessert. "We are pleased to say it is the custom with most intelligent planters of this section of our State to give their Negroes an annual barbecue," wrote an observer.⁵⁰ Barbecues were usually followed by music, dancing, and a baby show. Planters sometimes gave their Negroes a ration of whiskey at Christmas, and many provided a regular tobacco allowance for adult slaves, both male and female.⁵¹

Minimum clothing allowances for Negroes consisted of three suits and one pair of shoes a year, the latter being worn only during the winter. Boys under sixteen might require four suits of clothing a year. Women generally were provided with underclothing in the form of drawers, which were needed for outdoor wear during cold weather. Since rain-proof coats could not be bought cheaply, Negroes were furnished with a crude cape-like garment made of cotton osnaburgs well coated with boiled linseed oil.[52] This served as protection against both rain and wind.

When left to their own resources in providing clothing, Negroes generally were better dressed than the requirements of the plantation dictated. Their desire to possess good clothing and their tendency to imitate the dress of their master's family were said to have led to much petty thieving. Apparently to minimize this problem, the *Southern Cultivator* recommended that the costume of Negro slaves be regulated by law "like the peasantry of rural populations of other countries," and that Negroes not be allowed to array themselves in public in the cast-off finery of their masters and mistresses. "To a person of refined taste, the airs and assumptions of dandified Negroes . . . is most disgusting and offensive," it concluded.[53]

One indication of improved slave management in the 1850's was the increased attention generally given to such problems as the laundry of clothing and to maintaining cleanliness inside the cabin and about the quarters. "The negro is naturally inclined to be filthy," observed a Stewart County planter. "He should have at least every other Saturday afternoon to wash, and then be required to keep himself clean." He urged that slave quarters be inspected frequently and all filth removed from around them.[54] "Fleas in the cabin will make grass in the cornfield," ran the proverb, "for he that catches fleas by night will catch sleep by day."

Alleviation of malaria and the "summer sickness" was thought to result from whitewashing the inside of cabins, a task usually done during August. Frugality in the supply of firewood was criticized as false economy as were such practices as providing insufficient blankets. Cabins built too low on the ground with accumulated filth underneath, two families in the same house, lack of shade trees in the yard (which, however, must not be too close to the cabin), and improper and poorly located privies were also condemned. "Lean woodpiles make a fat graveyard; small cabins make large graveyards," and "much filth makes much physic," ran some of the proverbs on these problems.[55]

Ministers often preached sermons on the relations between master and slave, and such sermons occasionally appeared in pamphlet form. The Ohio-born father of Woodrow Wilson, in a pamphlet written on the subject in the 1850's, described the ideal relationship between master and slave as identical to that between parents and children. Thomas F. Scott of Columbus, in an essay on the management of slaves, emphasized the master's obligation to provide for the slave a mild and considerate form of justice, innocent amusements, ample food, shelter, and clothing, care in old age and infirmity, and moral and religious training. Charles W. Howard insisted that the term "slavery" had been erroneously used for "the proper and natural relation of the negro and white." He insisted that the word "serf" was more suggestive of the true relationship.[56] The term "servant" came to be widely substituted for "slave" in the waning decades of the plantation regime.

The wills of plantation masters often reflect a growing spirit of *noblesse oblige* during this period. "I wish my man, Grig, to be held as the nominal property of such master as he may choose but enjoy all the fruits of his labor or at his option to be manumitted and sent out of the State," was the direction given by Joel Crawford, who died in Early County in 1858. As a reward for their fidelity, several of Crawford's Negroes were allowed to select their future masters, though such practice might involve some diminution in price. In no instance were husbands to be separated from their wives, nor mothers from their young children. Aged and infirm Negroes were to receive a comfortable support from their master's estate.[57]

The attempt to improve working conditions of the plantation slave generally was motivated by economic as well as humane considerations, for improving his efficiency as a laborer promised greater profits to the owner. Originally plantation rules were established to insure maximum labor with little regard for the factors which promoted high efficiency. The Negro's inclination to sing while working and to move with the tempo of his song encouraged the teaching of lively tunes. The Negro spiritual may owe its origin in part to the slave's tendency to resist such speed-up contrivances designed by ambitious foremen. As a better understanding of the Negro developed, plantation rules became more flexible. While a system of rewards and punishments remained the basis of plantation discipline, the more successful masters came to respect the personality of the slave and to deal with him as an individual. "With one, perhaps, moral suasion will do—with another flattery, while with the majority the fear of punishment is the motive power," said a Stewart County farmer. This master permitted his adult slaves to have a plat of their own, to the planting and cultivation of which he gave his personal supervision. "I give my negroes as much money as I am willing to see thrown away in the indulgence of a childish fancy," he said.[58]

Slaves generally were required to get plenty of sleep. "They are thoughtless and if allowed to do so, will set on a stool or chair and nod or sleep till morning," said one. Marriage within the plantation community was suggested in order to minimize visiting, yet too strict adherence to this practice was found to result in idiotic children and other evils of inbreeding, and the practice was far from universal.[59]

Many planters recognized, as employers of free labor failed to do, that a law of diminishing returns made it highly advisable to limit working hours and conditions. A Georgian in 1849 tried Martin W. Philips' plan of allowing laborers and work animals two hours of rest at noon, increasing the period to three hours as the heat of summer became more intense. "Our crop was cultivated better than ever before, and the mules [were] in better condition, as were also the hands," he reported. James Thomas of Hancock County was a meticulous and painstaking experimenter. All of his arable land was carefully measured and staked off into squares of seventy yards each. He found that his Negroes would do fifteen per cent more work in very hot weather if given a five-minute rest period after each thirty minutes of active labor. "... [They] will quit work at night as fresh as they entered it in the morning and they will get home at night fresh and happy and ... demonstrate it by 'patting and dancing juber,'" he said, "whilst your neighbors' negroes will get slow by 12 o'clock, still slower by night, and if they get from the cotton field to the house, half the young negroes will drop to sleep without supper."[60] He prohibited Sunday labor and night work, and females were assigned lighter tasks, particularly during periods of pregnancy.[61]

One of the more significant contributions made in Georgia to the management of slaves was by David Dickson. Employing no overseers, Dickson personally trained his Negroes to become efficient and expert operatives. He claimed to have increased a worker's daily pick of cotton by a hundred pounds by teaching him to pull all of the cotton out of the boll with a single stroke and with the same motion to strike at the next one. Similar efficiency was stressed in handling the axe and hoe, and in the execution of other work. Dickson insisted on good tools and he did not begrudge the time spent in repairing and resharpening them. He did not find it necessary to maintain a constant vigilance over his workers, and visitors invariably commented upon their happy condition and on the equally fine appearance of his crops and livestock. His Negroes were said to possess a craftsman's pride in their work and to strive for high production. "The truth is," said one, "his hands see that they beat their neighbors, and people are constantly coming to see their fine crops, and they feel a pride in their success."[62]

Closely related to the changing attitudes toward slavery was a change in the managerial aspects of plantation economy. A Burke County planter in 1844 pointed out that an overseer's only qualifications were farming experience and sobriety, and the sole test of his management was the size of his crops. "But revolution is at hand," said he, in emphasizing the changing role of the overseer in a regime of improved cultural methods, with emphasis on soil-building, of breeding hardier livestock, and of better care and management of Negroes. He urged Georgians to turn to better trained overseers who could manage a diversified plantation.

Another planter believed that all overseers should be discharged because of their general inefficiency and lack of interest in the land, and that masters should learn to manage their own lands. The overseer belonged to an inarticulate group and he often was misinterpreted by those who wrote about him. Trained in the hard school of experience, he was poorly paid and his tenure was uncertain. Given maximum responsibility with minimum authority, his primary function was the supervision of slave laborers, and these had a much higher value than the land on which they worked. He might be dismissed for a slight abuse of the Negro, but seldom for the most wretched abuse of the land. Agricultural reformers tended to identify the overseer with absentee planters, lack of food crops, the abuse of land, and unscientific management of livestock.[63] The overseer, like the share-cropper of a later era, generally had little interest in agricultural reform, although the career of Garland D. Harmon shows him to have been an outstanding exception to such a generalization.

Between the large planters and the small farmers were many operators who employed managerial labor only in prosperous years. Hence the employment possibilities of overseers fluctuated somewhat with the price of cotton. The low price of cotton in the early 1840's resulted not only in the dismissal of many overseers, but caused some marginal farmers to forsake their businesses in search of employment as managers of large establishments. Opportunities for such employment, like the wage scale of overseers, failed to keep pace with advancing cotton prices during the 1850's.

Keen competition, short tenure, and the employer's lack of confidence in him were problems which the overseer faced, and these problems appeared to grow more acute as the plantation era approached the Civil War era. Planters were becoming more exacting in their requirements, some even recommending the establishment of normal schools for the training of overseers and the elevation of the overseership to the level of a dignified and honorable profession. The overseership never became an honored profession, however.[64] As masters became more solicitous for the welfare of their slaves, the Negro was not slow to push his advantage, and the overseer found his work losing some of its former dignity. "Everyone conversant with Negro character, knows well their proclivity for lying and stealing," complained one. "Make enquiry of them, and the owner can get a budget of news, sufficient to hang any overseer." He called his job as mean and contemptible as any which existed. "I am just tired of it, and will quit it, as soon as I can find a better business," he concluded.[65]

The reform movement probably resulted in better overseers on the cotton plantations, but the Georgia planter was often outbid for their services by growers in the Southwest. Garland D. Harmon was among the better overseers who moved westward with the expanding tide of cotton planting. Thomas Affleck, in 1855, said that efficient overseers were not to be had in some parts of the South. He added that there were many overseers in Mississippi and Louisiana who were well educated and in some instances much more competent than their employers. Some planters predicted that they would soon be compelled to pay as much as fifteen hundred dollars for the yearly services of overseers, when formerly five hundred dollars was considered excellent remuneration.[66]

There is little doubt that Southern slavery grew steadily more temperate in the waning years of its life. And unprejudiced observers found much to commend in the practice of the institution in Georgia in the 1850's by native-born Northerners transplanted to farm life in Georgia. Richard Peters, Daniel Lee, Charles A. Peabody, Jarvis Van Buren, and numerous others who came South adopted the system in its entirety. Daniel Lee was driven to outright secessionism by the anti-slavery propaganda of the Northern abolitionists.[67] Solon Robinson, the popular Northern agricultural commentator of the age, thought that slavery was one of the best institutions for the Negro that could be devised, and he reiterated his hatred for the abolitionists. Much of this pro-slavery sentiment may be attributed to the unifying forces of the general agricultural movement, which was largely non-sectional and even international in its broadest outlook, and often transcended political barriers.

Notes

Chapter I

1. Ellis Merton Coulter, ed., *The Journal of William Stephens, 1741-1743* (Athens, 1958), 183 (cited hereafter as Stephens' *Journal* (1741-1743); John Payne, *Georgia From Latest Authorities* (New York, 1799), 422.
2. James D. B. De Bow, *The Industrial Resources, Statistics, Etc., of The United States . . .*, 3 vols. (New York, 1854), I, 355; Allen D. Candler and Lucian Lamar Knight, eds., *The Colonial Records of the State of Georgia*, 26 vols. (Atlanta, 1904-1916), XIX, Part II, 398 (cited hereafter as *Colonial Records*); Eliza Lucas (Mrs. E. L. Pinckney), *Journal and Letters of Eliza Lucas* (Wormsloe, 1850), 28.
3. William Bartram, *Travels Through North and South Carolina and Georgia, East and West Florida . . .* (Philadelphia, 1791), 23; Payne, *op. cit.*, 443.
4. Patrick Tailfer, et al., *A True and Historical Narrative of the Colony of Georgia, in America, From the First Settlement Thereof until this present Period . . .* (Charleston, 1741), 29, 101-03; *Colonial Records*, II, 337.
5. Ellis Merton Coulter and Albert B. Saye, eds., *A List of the Early Settlers of Georgia* (Athens, 1949), 2 et passim.
6. Alexander R. MacDonel, "The Settlement of the Scotch Highlanders at Darien," *Georgia Historical Quarterly*, XX (Sept. 1936), 252-53.
7. *Colonial Records*, II, 164-67; XXII, Part II, 93.
8. *Ibid.*, II, 275-82.
9. *Ibid.*, 407.
10. *Ibid.*, 495.
11. "A List of Salzburgers Shipped on Board the Loyal Judith . . .," September, 1744," no. 14212, p. 69, in Phillips Collection, University of Georgia Library.
12. *Ibid.*, 97; *An Account Shewing the Progress of the Colony of Georgia in America from its First Establishment* (London, 1741), 61, 62, in De Renne Collection, University of Georgia Library.
13. Thomas Stephens, *The Hard Case of the Distressed People of Georgia* (London, 1742), 2.
14. *Colonial Records*, XXIV, 436.
15. *An Account Shewing the Progress of the Colony of Georgia . . .*, 15.
16. *Ibid.*, 33, 64-65.
17. *Colonial Records*, I, 56-62, 506-07; IV, 523; XIII, 160; XXV, 225, 290, 347-51; Tailfer, *op. cit.*, 114.
18. See *Colonial Records*, VI-XI.
19. *Colonial Records*, III, 373-80, 404-13.
20. Milton Sydney Heath, *Constitutional Liberalism: The Role of the State in Economic Development in Georgia*

to 1860 (Cambridge, 1954), 35-37.
21. *An Account Shewing the Progress of the Colony of Georgia* . . ., 43.
22. *Ibid.*, 5.
23. Tailfer, *op. cit.*, 29; Hesta Walton Newton, "The Agricultural Activities of the Salzburgers in Colonial Georgia," *Georgia Historical Quarterly*, XVIII (Sept. 1934), 251.
24. William Harden, *A History of Savannah and South Georgia*, 2 vols. (Chicago, 1913), I, 20; William B. Stevens, *A History of Georgia*, 2 vols. (New York, 1847), I, 99.
25. Tailfer, *op. cit.*, 29.
26. Stephens' *Journal* (1741-1743), 155.
27. *Colonial Records*, II, 377.
28. Stephens' *Journal* (1741-1743), 152.
29. Plat Book "C," Surveyor General's Office, 22, in Georgia Department of Archives and History.
30. *An Account Shewing the Progress of the Colony of Georgia* . . ., 61, 62.
31. Thomas Stephens [?], *A Brief Account of the Causes that have Retarded the Progress of the Colony of Georgia, in America* . . . (London, 1743), 13-14 (cited hereafter as *A Brief Account*).
32. *An Account Shewing the Progress of the Colony of Georgia* . . ., 33; *Colonial Records*, II, 394.
33. *Colonial Records*, II, 500.
34. Heath, *op. cit.*, 63, 64.
35. *Ibid.*, 70.
36. *Ibid.*, 57.
37. Evarts B. Greene and Virginia D. Harrington, *American Population Before the Federal Census of 1790* (New York, 1932), 180-182.
38. *Colonial Records*, V, 653; VII, 457.
39. *Ibid.*, X, 754.
40. Heath, *op. cit.*, 64. See also "Records of land grants," *Colonial Records*, VI-XI; John G. W. DeBrahm, *History of the Province of Georgia* (Wormsloe, 1849), 51; "Letters from Sir James Wright," *Collections of the Georgia Historical Society*, 11 vols. (Savannah, 1840-1955), VI, 102 (cited hereafter as *Collections*).
41. James Wright, *A Proclamation* (Savannah, June 11, 1773), in De Renne Collection, University of Georgia Library.
42. Curtis P. Nettles, *The Roots of American Civilization* (New York, 1938), 609-10.
43. Heath, *op.cit.*, 69.
44. *Ibid.*, 70.
45. *Colonial Records*, XIX, Part II, 152-62; Allen D. Candler, ed., *The Revolutionary Records of the State of Georgia*, 3 vols. (Atlanta, 1908), I, 326-48, 373-97, 611; III, 386, 365, 570 (cited hereafter as *Revolutionary Records*).
46. *Revolutionary Records*, I, 279-80.
47. *Colonial Records*, XIX, Part II, 53-58.
48. *Revolutionary Records*, II, 73-74, 385.

CHAPTER II

1. Heath, *op. cit.*, 29.
2. Harden, *op. cit.*, I, 21.
3. DeBrahm, *op. cit.*, 43, 53; Stephens' *Journal* (1741-1743), 188, 194; William Stephens, *A Journal of the Proceedings in Georgia* [Oct. 20, 1737-Oct. 4, 1740], in *Colonial Records*, IV, 522 (cited hereafter as Stephens' *Journal* (1737-1740); *Colonial Records*, V, 696.
4. Hector B. Beaufain, Purysburg, S. C., to the Earl of Egmont, Dec. 12, 1741, Phillips Collection, University of Georgia Library, No. 14212, pp. 105-13; *Colonial Records*, II, 499-500.
5. Ellis Merton Coulter, ed., *The Journal of William Stephens, 1743-1745* (Athens, 1959), 80 (cited hereafter as Stephens' *Journal* (1743-1745).
6. Stephens' *Journal* (1737-1740), 310.
7. Thomas Boreman, *A Compendious Account of the Whole Act of Breeding, Nursing and Right Ordering of*

the Silkworm (London, 1733), 14-24.
8. *Ibid.*, 22; Stephens' *Journal* (1741-1743), 70.
9. Boreman, *op. cit.*, 21-24.
10. *Ibid.*, 21-24; William Stephens, *A Journal of the Proceedings in Georgia* [Oct. 5, 1740-Oct. 28, 1741]. *Colonial Records*, Supplement to Vol. IV, 141-42 (cited hereinafter as Stephens' *Journal* (1740-1741).
11. Boreman, *op. cit.*, 12-24.
12. *Colonial Records*, II, 480, 482.
13. *Ibid.*, 491, 519.
14. Stephens' *Journal* (1741-1743), 168-69.
15. DeBrahm, *op. cit.*, 22.
16. J. M. Hofer, "The Georgia Salzburgers," *Georgia Historical Quarterly*, XVIII (June 1734), 107.
17. Stephens' *Journal* (1741-1743), 62.
18. Stephens' *Journal* (1743-1745), 156.
19. Joseph Scott, *The United States Gazetteer* (Philadelphia, 1795), n.p.
20. DeBrahm, *op. cit.*, 50.
21. F. Hoff, *Agricultural and Commercial Almanac Calculated for the States of Georgia and South Carolina for the Year 1807* . . . (Charleston, 1807), n.p.
22. Alexander Hewat, *An Historical Account of the Rise and Progress of South Carolina and Georgia*, 2 vols. (London, 1779), II, 140-41.
23. Hennig Cohen, "A Colonial Poem on Indigo Culture," *Agricultural History*, XXX (January 1956), 41; Eliza Lucas (Mrs. E. L. Pinckney), *Journal and Letters of Eliza Lucas* (Wormsloe, 1850), 5.
24. *Colonial Records*, I, 362; XXV, 55.
25. *Colonial Records*, XXV, 251; Stephens' *Journal* (1741-1743), 201.
26. *Colonial Records*, XXV, 55.
27. D. B. Worden, *A Statistical Account of the United States of North America From the Period of Their First Colonization to the Present Day*, 2 vols. (Edinburgh, 1819), 482.
28. Hewat, *op. cit.*, II, 143-44; *Colonial Records* XIX, Part II, 33.
29. Hewat, *op. cit.*, II, 140-46; Letter Book, Clay Telfair and Co., Vol II, July 1765 (Ms. in the Georgia Historical Society Library, Savannah).
30. *The Georgia Gazette* (Savannah), Feb. 17, 1765; DeBrahm, *op. cit.*, 53; *Acts Passed by the General Assembly of Georgia From Feb. 17, 1755 to May 10, 1789* (Savannah and Augusta), 455-56 (cited hereafter as *Acts of the General Assembly, 1755-1789*).
31. *Acts of the General Assembly, 1755-1789*, 455-56; *Colonial Records*, II, 437.
32. Tailfer, *op. cit.*, 102-03; Beaufain to Egmont, *loc. cit.*, 110-11.
33. *Ibid.*, 110.
34. Stephens' *Journal* (1737-1740), 190; Thomas Stephens [?], *A Brief Account* . . ., 9-10 (appendix).
35. Stephens' *Journal* (1737-1740), 256.
36. Stephens' *A Brief Account* . . ., 10-11; Stephens' *Journal* (1740-1741), 121, 125.
37. Stephens' *Journal* (1737-1740), 542, 663, 670.
38. Tailfer, *op. cit.*, 102, 103.
39. *Ibid.*, 30.
40. Stephens, *A Brief Account* . . ., 40-41.
41. Hester W. Newton, "The Agricultural Activities of the Salzburgers in Colonial Georgia," *Georgia Historical Quarterly*, XVIII, (Sept. 1934), 251.
42. *An Account Shewing the Progress of the Colony of Georgia* . . ., 66-67.
43. *Ibid.*, 35-36.
44. *Colonial Records*, V, 372.
45. *Ibid.*, II, 437.
46. *Ibid.*, XXIII, 470.
47. Stephens' *Journal* (1741-1743), 138.
48. *Ibid.*, (1737-1740), 680.
49. *Colonial Records*, XXIV, 358-62.
50. *Ibid.*, XXIV, 437.
51. Stephens' *Journal* (1743-1745), 210.
52. Book "D," Estates and Inventories, 116 (in Georgia Department of Archives and History).
53. DeBrahm, *op cit.*, 53.
54. Letter Book, Joseph Clay and Co., Vol. I, Dec. 19, 1772, to March 31, 1774.
55. Hewat, *op. cit.*, I, 267.
56. *Acts of the General Assembly, 1755-1789*, 344, 429.

57. Hewat, *op. cit.*, I, 96-97.
58. Thomas Stephens, *The Method and Plain Process For Making Pot-Ash Equal if not Superior To the Best Foreign Pot-Ash* (London, 1775), 1-12.
59. S. Urlsperger, *Ausfahrliche Nachricten vondem Salzburgischen Emigration* (Halle, 1735-1752), I, III; Stephens' *Journal* (1737-1740), 286.
60. Stephens' *Journal* (1741-1743), 178; *Ibid.*, (1737-1740), 25; *An Account Shewing the Progress of the Colony of Georgia* . . ., 35-36; *Colonial Records*, V, 658.
61. *Ibid.*, V, 513.
62. *Ibid.*, V, 172, 247, 395, 413, 502, 705, 722.
63. Stephens' *Journal* (1741-1743), 58, 178.
64. *Colonial Records*, II, 275.
65. *Ibid.*, V, 248.
66. Stephens' *Journal* (1737-1749), 314.
67. John H. Logan, *A History of the Upper Country of South Carolina From the Earliest Periods to the Close of the War of Independence*, 2 vols. (Columbia, 1859), I, 152-53.
68. Stephens' *Journal* (1740-1741), 90.
69. *Ibid.*, (1741-1743), 175.
70. *Ibid.*, (1737-1740), 286.
71. *Colonial Records*, XXIV, 362.
72. Stephens' *Journal* (1741-1743), 58, 69, 183.
73. *Colonial Records*, V, 584.
74. Stephens' *Journal* (1741-1743), 190.
75. *Colonial Records*, XXIV, 362.
76. Stephens' *Journal* (1737-1740), 300.
77. *Colonial Records*, XXII, Part II, 169.
78. *Ibid.*, 171.
79. Beaufain to Egmont, *loc. cit.*, 93; Stephens' *Journal* (1737-1740), 314.
80. *Ibid.*, (1731-1743), 208.
81. *Ibid.*, (1737-1740), 286.
82. Marks and Brands, 1755-1793 (in Georgia Department of Archives and History).
83. *The Georgia Gazette*, Feb. 17, 1765.
84. Stephens' *Journal* (1737-1740), 513.
85. *Ibid.*, (1741-1743), 77.
86. Plat Book "B," Surveyor General's office; *Colonial Records*, V, 661; XXIV, 362; Stephens' *Journal* (1741-1743), 77, 214.
87. Plantation Account Book of Richard Leake, 1785-1801 (a Ms. in the Georgia Historical Society Library).
88. Stephens' *Journal* (1740-1741), 16.
89. *Colonial Records*, II, 507.
90. DeBrahm, *op. cit.*, 53; *Acts of the General Assembly, 1755 to 1789*, p. 429.
91. *Ibid.*, 344, 455-56.
92. Thomas Stephens, *A Brief Account* . . ., 1-3 (appendix).
93. *Colonial Records*, VII, 831.
94. Stephens' *Journal* (1740-1741), 207.
95. *Southern Cultivator*, II (June 1844), 106.
96. Frances Moore, *A Voyage to Georgia Begun in the Year 1735* (London, 1744), Georgia Historical Society, *Collections*, I, 115-16.
97. P. A. Strobel, *The Salzburgers and Their Descendants* (Baltimore, 1855), 7; Stephens' *Journal* (1743-1745), 176.
98. Caleb Swann, "Position and State of Manners and Arts in the Creek or Muscogee Nation in 1791," Henry R. Schoolcraft, ed., *Information Respecting . . . the Indian Tribes of the United States*, 6 vols. (Philadelphia, 1855), V, 256.

CHAPTER III

1. Heath, *op. cit.*, 93.
2. *Ibid.*, 95; S. Guyton McLendon, *History of the Public Domain of Georgia* (Atlanta, 1924), 45 (cited hereafter as *Public Domain*).
3. *Revolutionary Records*, II, 605-06; Robert and George Watkins, *Digest of the Laws of the State of Georgia . . . to 1798* (Philadelphia, 1800), 289 (cited hereafter as Watkins, *Digest*).
4. *Colonial Records*, XIX, Part II, 302-04; *Revolutionary Records*, II, 225, 237-43.

5. William B. Stevens, *A History of Georgia*, 2 vols. (New York, 1847), II, 356-58.
6. *Revolutionary Records*, II, 787-99.
7. W. B. Stevens, *op. cit.*, II, 354, Plat Book "B," Surveyor General's Office.
8. *Revolutionary Records*, III, 267, 278-80, 388-89; "Letters of Joseph Clay, of Savannah, Georgia, 1776-1793," Georgia Historical Society, *Collections*, VIII, 171.
9. *Ibid.*, 47-212; Reba C. Strickland, *Religion and the State of Georgia in the Eighteenth Century* (New York, 1939), 162-63.
10. *Revolutionary Records*, II, 556.
11. *Colonial Records*, XIX, Part II, 516-17; Watkins, *Digest*, 422, 438, 474, 515, 571, 615.
12. *Revolutionary Records*, III, 359-60.
13. S. G. McLendon, *Public Domain*, 46-58.
14. *Ibid.*, 175; Absalom H. Chappell, *Miscellanies of Georgia, Historical, Biographical, Descriptive, Etc.* (Atlanta, 1874), 51 (cited hereafter as *Miscellanies of Georgia*).
15. S. G. McLendon, *Public Domain*, 58.
16. Charles H. Haskins, "The Yazoo Land Companies," *Papers of the American Historical Association* (New York, 1891), V, 80, 103, Watkins, *Digest*, 530; S. G. McLendon, *Public Domain*, III, VIII, XI.
17. S. G. McLendon, *Public Domain*, 183-200; E. M. Coulter, *A Short History of Georgia* (Chapel Hill, 1933), 188.
18. S. G. McLendon, *Public Domain*, 144-50; Charles J. Kappler, *Indian Affairs, Laws and Treaties*, 3 vols. (Washington, 1892-1913), II, 85-86.
19. *Senate Journal* (Louisville, 1803), 4.
20. *The Augusta Chronicle*, Dec. 4, 1802.
21. Harriet Milledge Salley, ed., *Correspondence of Governor John Milledge* (Columbia, S.C., 1949), 64-69.
22. Heath, *op. cit.*, 141.
23. Oliver H. Prince, *Digest of the Laws of the State of Georgia* (Athens, 1837), 545 (cited hereafter as Prince's *Digest*); *Senate Journal* (Louisville, 1804), 68-69.
24. A. H. Chappell, *Miscellanies of Georgia*, 26; S. G. McLendon, *Public Domain*, 121-26.
25. See records of lottery grants in the office of the Surveyor General, Georgia Department of Archives and History. See also the following published records of lottery grants: Martha Lou Houston, *Land Lottery List of Oglethorpe County, Georgia, 1804, Hancock County Georgia, 1806* ... (Columbus, 1928); and *Reprint of the Official Register of Land Lottery of Georgia, 1827* (Columbus, 1928); James F. Smith, *The Cherokee Land Lottery* (New York, 1938); *Gold and Land Lottery Register* (printed for the state at Milledgeville in 1833).
26. These estimates are based on figures from Benjamin H. Hibbard, *A History of the Public Land Policies* (New York, 1939), 101-15.
27. Records of Surveyor General's office.
28. Based on estimates from Heath, *op. cit.*, 157.
29. Jedidiah Morse, *A Report to the Secretary of War of the United States on Judian Affairs* (New Haven, 1822), 146.
30. See James C. Bonner, "Tustunugee Hutkee and Creek Factionalism on the Georgia-Alabama Frontier," *The Alabama Review*, X (April 1957), 111 et passim.
31. The Treaty of New York in 1790 was particularly obnoxious to Georgians. What angered them most was the acceptance of the line of the Oconee River as the state's western boundary and the implied intention of the federal government that Indians residing west of this line would be transformed from hunters into farmers. To achieve the change the United States agreed to furnish them such commodities as horses, tools, and seed.
32. Ulrich B. Phillips, *Georgia and State Rights* (Washington, 1902), 140.
33. *Hancock Advertiser* (Mt. Zion), August 12, 1828; Ulrich B. Phillips,

NOTES 209

History of Transportation in the Eastern Cotton Belt to 1860 (New York, 1908), 125.
34. John T. Henderson, *The Commonwealth of Georgia* (Atlanta, 1884), 311, 312.
35. *Fifth Census of the United States* (Washington, 1832), 14-15; *Compendium of the Sixth Census* (Washington, 1841), 49-51.
36. *Fifth Census of the United States* (1830), 96-97.
37. *Compendium of the Sixth Census*, 202; *Seventh Census of the United States* (Washington, 1853), 367; Free Inhabitants of Hancock County in 1850, Schedule III (a microfilm of the original manuscript of the Census of 1850 in the Woman's College of Georgia Library); Free Inhabitants of Carroll County in 1850, Schedule III, *loc. cit.; Hancock Advertiser*, Dec. 14, 1829, Jan 4, 1830; *Niles' Weekly Register*, XXVI (Jan. 1830), 313 (cited hereafter as *Niles' Register*).
38. *Hancock Advertiser*, Feb. 25, 1828.
39. Mrs. Anne (Newport) Royall, *Mrs. Royall's Southern Tour, or Second Series of the Black Book*, 2 vols. (Washington, 1831), II, 117, 138.
40. George W. Featherstonehaugh, *Excursion Through the Slave States*, 2 vols. (London, 1844), II, 320.
41. *Niles' Register*, XXXIV (June 1828), 248; *Hancock Advertiser*, June 28, 1828.
42. A fairly complete file of land titles and transactions for Carroll County is available in the office of the Clerk of Superior Court. A dossier showing abstract of titles by lot number may be found in the law office of Boykin and Boykin, Carrollton, Georgia.
43. *Compendium of the Sixth Census* (1840), p. 202; *Acts of the General Assembly* (1853-1854), 73.
44. James D. B. DeBow, *Industrial Resources*, I, 355; James C. Hemphill, *Climate, Soil, and Agricultural Capabilities of South Carolina and Georgia* (Washington, 1882), 47-54; *Southern Cultivator*, XVI (Oct. 1858), 305.
45. Ulrich B. Phillips, et al., eds., *A Documentary History of American Industrial Society*, 10 vols. (Cleveland, 1910), II, 167.
46. Emily P. Burke, *Reminiscences of Georgia* (Oberlin, 1850), 23, 24, 209.
47. Richard Keily, *A Brief Description and Statistical Sketch of Georgia* (London, 1849), 25-26.
48. Emily Burke, *op. cit.*, 209.
49. Rufus Anderson, *Memoir of Catherine Brown* (New York, 1827), 1 *et passim;* Morse, *op. cit.*, 152, 169; *Niles' Register*, XXXIX (Sept. 1830), 992; John T. Henderson, *op. cit.*, 311, 312.
50. Henry T. Malone, *Cherokees of the Old South: A People in Transition* (Athens, 1956), 179-84.
51. *Fifth Census of the United States* (1830), 96, 97.
52. Charles H. Smith, *The Farm and Fireside* (Atlanta, 1891), 8-11.
53. Hiram P. Bell, *Men and Things* (Atlanta, 1907), 6-7, 10.
54. *Hancock Advertiser*, July 19, 1830.
55. Charles Lanman, *Letters From the Allegheny Mountains* (New York, 1849), 20-24, 47.
56. Productions of Agriculture in Habersham County in 1860, Schedule IV (a microfilm of the original manuscript of the Census of 1860 in the Woman's College of Georgia Library).
57. Minute Book "I," Records of Habersham County, Ordinary's office.
58. As early as 1789 a steamboat invented by William Longstreet appeared on the Savannah River. In 1819 the first steamboat reached Milledgeville from Darien. Ten years later steamboats had reached Macon on the Ocmulgee and Columbus on the Chattahoochee.

CHAPTER IV

1. Thomas Spalding, "Union Agricultural Society of Georgia," *American Farmer*, VII (Sept. 2, 1825), 185-87.
2. Letter Book "B," Clay, Telfair and Co., Vol. 2 (May 22, 1785 to Feb. 20, 1786), Feb. 20, 1786.
3. *Ibid.*, June 11, 1785; Harden, *op. cit.*, I, 271; *American Farmer*, VII (Sept. 2, 1825), 185-87.
4. Warden, *op. cit.*, 483.
5. Albert V. House, "Labor Management Problems on Georgia Rice Plantations," *Agricultural History*, XXVIII (Oct. 1954), 149 *et passim*.
6. Salley, *op. cit.*, 159.
7. Rice continued to remain an important item in the diet of people living along the Georgia coast. Even by the mid-twentieth century inventories of coastal merchants showed relatively large amounts of this item.
8. *The American Museum, or, Universal Magazine* (Philadelphia), March 1792, p. 124.
9. J. F. D. Smyth, *A Tour in the United States of America*, 2 vols. (London, 1784), II, 40; William Lee, *The True and Interesting Travels of William Lee . . .*, (New York, 1782), 22-23.
10. John Melish, *Travels Through the United States of America in the Years 1806 and 1807 and 1811*, 2 vols. (Philadelphia, 1815), I, 27, 574.
11. Worden, *op. cit.*, I, 482.
12. Lewis Cecil Gray, *History of Agriculture in the Southern United States to 1860*, 2 vols. (New York, 1941), II, 748.
13. Letter Book, Clay, Telfair and Co., Vol. 1, April 27, 1784.
14. *Ibid.*, Vol. 2, June 25, 1785.
15. Of particular significance are the farm and plantation inventories covering this period which are available in the Ordinary's records of Richmond, Jefferson, Greene, and Hancock counties. See Richmond County, Book "A" (Inventories 1799-1813); Jefferson County, Book "C" (Inventories, 1816-1823); Greene County, Book "E" (Estate Records, 1795-1809). Additional Greene County records for this period are to be found in Book "F" and Book "K." The Ordinary's records on the administration of estates in Hancock County cover the entire period beginning in 1793.
16. George Sibbald, *Notes and Observations on the Pine Lands of Georgia . . .* (Augusta, 1801), 62.
17. Book "K," Records of Greene County, Court of Ordinary, June, 1806, pp. 243-53.
18. Harden, *op. cit.*, I, 21; *American Farmer*, VII (Sept. 2, 1825).
19. Book "A," Records of Richmond County, Court of Ordinary, 41.
20. Book "F," Records of Greene County, Court of Ordinary, 70-73.
21. Stephens' *Journal* (1737-1740), 541.
22. *Colonial Records*, V, 180.
23. Stephens' *Journal* (1737-1740), 541; *Colonial Records*, XXII, part II, 351.
24. Melish, *op. cit.*, I, 27; *Southern Cultivator*, II (May 1844), 83.
25. *Ibid.*, X (June 1852), 184; *The Country Gentleman*, IX (Feb. 19, 1857), 147.
26. *Southern Cultivator*, II (May 1844), 83.
27. *Ibid.*; F. Hoff, *Agricultural and Commercial Almanac . . .*, n.p.
28. Payne, *op. cit.*, 444.
29. Denison Olmsted, *Memoir of Eli Whitney* (New Haven, 1846), 13; *Hunt's Merchants' Magazine*, II (March 1841), 212.
30. Olmsted, *Memoir of Eli Whitney*, 20.
31. *Ibid.*, 13, 32.
32. L. C. Gray, *op. cit.*, II, 681.
33. William Lee, *True and Interesting Travels . . .*, 22-23.
34. *Southern Cultivator*, XV (July 1857), 207. See also inventories of farm estates for Richmond, Jefferson, Greene, and Hancock counties, *loc. cit.*
35. This term was first used in Jefferson County in 1818. See Jefferson County Ordinary's records, Book "C" (Inven-

NOTES 211

tories, 1816-1823), p. 155.
36. *Southern Cultivator,* XV (July 1857), 207-09.
37. *Niles' Register,* XV (Oct. 1818), 135; Phillips, *A History of Transportation* . . ., 237.
38. Albert H. Sanford, *The Story of Agriculture in the United States* (New York, 1916), 135-43; *Southern Cultivator,* III (July 1845), 99-100.
39. *Ibid;* Hiram P. Bell, *Men and Things* (Atlanta, 1907), 8-9. A few simple threshing machines capable of cleaning sixty bushels of wheat in an hour were in use in Virginia as early as 1800. John Nicholson, *The Farmer's Assistant* (Richmond, 1820), 379.
40. *American Farmer,* IV (Oct. 4, 1822), 218.
41. David W. Lewis, ed., *Transactions of the Southern Central Agricultural Society, 1846-1851* (Macon, 1852), 378-80.
42. *Southern Agriculturist,* II (Feb. 1829), 65; (May 1829), 215.
43. *Hancock Advertiser,* Sept. 27, 1829; *American Farmer,* X (Dec. 1828), 324-25.
44. *Ibid.,* II (Dec. 15, 1820), 304.
45. *Southern Agriculturist,* II (Feb. 1829), 65.
46. James D. B. DeBow, *The Commercial Review of the South and West,* III (May 1847), 444 (cited hereafter as *DeBow's Review*).
47. Ulrich B. Phillips, *Life and Labor in the Old South* (Boston, 1929), 134-35; Burke, *Reminiscences* . . ., 223.
48. *Ibid.,* 177.
49. *Hancock Advertiser,* Sept. 14, 1829.
50. *Niles' Register,* XXXII (Aug. 1827), 372.
51. *Farmer's Register,* I (Sept. 1834), 384; *Southern Agriculturist,* II (July 1829), 318; *Hancock Advertiser,* Sept. 14, 1829.
52. *Ibid.,* May 24, 1830; *American Farmer,* XI (April 1829), 39, 45.
53. For weather and crop conditions in Georgia for these years, see *Niles' Register,* XXXII (August 1827), 372;

XXIV (June 1828), 245; XLIII (Sept. 1832), 20; XXVI (July 1830), 400; *American Farmer,* V (Oct. 1823), 212; IX (March 1827), 164; X (Jan. 1828), 52; *Southern Agriculturist,* II (July 1828), 318; *Hancock Advertiser,* Sept. 14, 1828; Anne Royall, *Southern Tour,* 138.
54. *The Georgia Journal* (Macon), quoted in *Hancock Advertiser,* April 14, 1828.
55. *Southern Agriculturist,* I (June 1828), 266.
56. *Niles' Register* XXIX (November 1825), 176; *American Cotton Planter,* II (Nov. 1829), 172, 283.
57. *American Farmer,* VII (April 1828), 34-35; *Georgia Journal,* Feb. 5, 1828.
58. *American Farmer,* VII (April 1828), 35.
59. *Hancock Advertiser,* March 10, 1828.
60. *Ibid.,* Apr. 14, 1828; *American Farmer,* VII (April 1828), 35, 36.
61. *Niles' Register,* XXXIII (Jan. 1828), 321.
62. *American Farmer,* X (Dec. 1828), 324-25; *Southern Agriculturist,* III (Nov. 1830), 561.
63. *Niles' Register,* XXV (Nov. 1828), 261.
64. *Ibid.,* 294; XXXVI (July 1829), 299. At militia musters during the following year, Georgia militiamen were clothed in homespun and equipped with "anti-tariff wooden muskets" made of poplar and soft pine. *Ibid.,* XXXVI (July 1829), 299.
65. *American Farmer,* XI (April 1829), 30.
66. *Niles' Register,* XXIV (Sept. 1828), 63.
67. Quoted in *Ibid.,* XXXV (Oct. 1828), 83.
68. *Ibid.,* XXXVI (May 1829), 162.
69. See Spalding's essays in *Southern Agriculturist,* I (March 1828), 106; (Oct. 1828), 433; (Dec. 1828), 522; II (Feb. 1829), 55, 57, 100, 103; (June 1829), 312, 347; (July 1829), 392; (Dec. 1829), 557; III (Feb. 1830), 71; (April 1830), 185, 186; (June 1830), 247, 293 (July 1830),

359, 556; (Oct. 1830), 405; (Nov. 1830), 517.
70. *Farmer's Register*, I (Jan. 1834), 490.
71. Bell, *Men and Things*, 6-7; Charles H. Smith, *The Farm and Fireside* (Atlanta, 1891), 8-11.
72. *Southern Cultivator*, II (April 1844), 49, 50; *Hancock Advertiser*, Sept. 1, 1828.

Chapter V

1. U. B. Phillips, *A History of Transportation* . . ., 70.
2. *Farmer's Register*, I (Jan. 1834), 490.
3. Herbert A. Keller, ed., *Solon Robinson, Pioneer and Agriculturist*, 2 vols. (Indianapolis, 1936), II, 378, 479.
4. *Southern Cultivator*, VIII (June 1850), 180.
5. Sir Charles Lyell, *A Second Visit to the United States of North America*, 2 vols. (New York, 1849), II, 25-26. Cited hereafter as *Second Visit*.
6. *Southern Agriculturist*, I (July 1828), 307.
7. *Hancock Advertiser*, March 3, 1828.
8. *Southern Cultivator*, V (Nov. 1847), 170; *Acts of the General Assembly of the State of Georgia* (Milledgeville: published annually by state printers, 1825-1841, and biennially from 1843-1853), 1840, appendix, 39. Cited hereafter as *Acts of the General Assembly*.
9. *Soil of the South*, II (Nov. 1852), 361; *Southern Cultivator*, II (June 1844), 92.
10. *Southern Agriculturist*, III (August 1830), 407.
11. DeBow, *Industrial Resources* . . ., 355.
12. U. B. Phillips, *A Documentary History* . . ., I, 131. Of more than 87,000 people in Florida in 1850, over 11,000 were natives of Georgia. There were at this time nearly 59,000 natives of Georgia in Alabama and 17,000 in Mississippi. R. Marsh Smith, "Migrations of Georgians to Texas," *Georgia Historical Quarterly*, XX (Dec. 1936), 322.
13. *Southern Cultivator*, II (April 1844), 92.
14. *Ibid.*, 9.
15. Quoted in *Ibid.*, XVII (March 1859), 76.
16. The census data for these counties indicate a heavy dominance of males as well as youth in the early population.
17. *Southern Cultivator*, III (July 1845), 100.
18. Sir Charles Lyell, *Second Visit*, II, 36.
19. *Southern Cultivator*, XVII (March 1859), 83.
20. *Soil of the South*, I (March 1851), 36.
21. *Ibid.*, 49.
22. *Southern Cultivator*, IX (Sept. 1851), 137.
23. *Southern Cultivator*, VIII (Oct. 1850), 307; XVII (Sept. 1859), 275.
24. *Ibid.*, XI (March 1853), 92.
25. *Ibid.*, XI (Feb. 1853), 78.
26. The counties of Walker, Gordon, Paulding, Gilmer, Cherokee, Cobb, Forsyth, Lumpkin, and Union in North Georgia also had fewer people in 1860 than ten years earlier. Some changes in county boundaries during the decade only partly explain these population losses.
27. *Soil of the South*, II (July 1852), 292.
28. *Genesee Farmer*, X (April 1849), 82.
29. Sir Charles Lyell, *Second Visit*, II, 25-26.
30. U. B. Phillips, *Life and Labor in the Old South*, 177-78.
31. John T. Henderson, *The Commonwealth of Georgia*, 62-63; Hiram P. Bell, *Men and Things*, 8.
32. *Southern Cultivator*, I (June 1843), 101.
33. Henry S. Randall, *Sheep Husbandry in the South* (Philadelphia, 1848), 60.
34. *Southern Cultivator*, XIV (Oct. 1856), 300.

35. *Soil of the South*, I (March 1851), 53.
36. See James C. Bonner "Profile of a Late Ante-Bellum Community," *American Historical Review*, XLIX (July 1944), 675.
37. *Soil of the South*, II (April 1852), 253.
38. Emily P. Burke, *op. cit.*, 118.
39. James Stuart, *Three Years in North America*, 2 vols. (New York, 1833), I, 65.
40. *Soil of the South*, I (Aug. 1851), 124-25.
41. William Lumpkin, Athens, Georgia, to his son at Holly Springs, Mississippi, Nov. 15, 1837 (in private possession).
42. *Ibid.;* Tyrone Power, *Impressions of America During the Year, 1833-1834, and 1835*, 2 vols. (London, 1836), II, 123.
43. George W. Featherstonehaugh, *Excursions . . .*, II, 319.
44. Nellie Peters Black, *Richard Peters, His Ancestors and Descendants* (Atlanta, 1904), 16, 18.
45. Sir Charles Lyell, *Second Visit*, II, 26-28; W. H. Sparks, *The Memories of Fifty Years* (Philadelphia, 1870), 20.
46. *Ibid.*, 28.
47. Andrew J. Prior, Rusk County, Texas, to George W. West, Cedartown, Georgia, Jan. 28, 1852, in the Prior-West Collection of the Duke University Library.
48. L. A. Prior, Sabine County, Texas, to George W. West, Dec. 4, 1853, in the Prior-West Collection, *loc. cit.*
49. R. Marsh Smith, "Migrations of Georgians to Texas," *loc. cit.*, 322.
50. Mary Shepperson Crabbe, Charles Alfred Peabody (an unpublished manuscript in the University of North Carolina Library), 1 *et passim*.
51. *Federal Union*, Jan. 17, 1860.
52. R. M. Smith, "Migrations of Georgians to Texas," *loc. cit.*, 322.
53. *Southern Cultivator*, XVIII (Dec. 1860), 384.
54. *American Cotton Planter and Soil of the South*, III (Jan. 1859), 18 (cited hereafter as *American Cotton Planter*).
55. *Soil of the South*, III (Jan. 1853) 394.
56. U. B. Phillips, *A Documentary History . . .*, II, 252-55.
57. *Seventh Census of the United States, 1850; population of the United States in 1860*.

Chapter VI

1. U. B. Phillips, *A History of Transportation . . .*, 48.
2. *Southern Agriculturist*, I (July 1828), 302; (August 1828), 337-85; *Niles' Register*, XXXIV (July 1825), 297; *American Farmer*, X (June 1829), 384; *Farmer's Register*, I (June 1834), 183; *DeBow's Review*, III (March 1847), 267.
3. *Hancock Advertiser*, April 28, 1826.
4. *Southern Agriculturist*, III (March 1830), 121.
5. *Ibid.*, 121, 122.
6. *Farmer's Register*, V (June 1837), 70-71.
7. *The Cultivator* [Albany, New York], IX (Nov. 1842), 187. Cited hereafter as *Albany Cultivator*.
8. *Ibid.*, VII (July 1840), 106.
9. *DeBow's Review*, III (May 1847), 444.
10. *Southern Cultivator*, I (March 1843), 82.
11. In 1851 Meriwether County farmers won the greater portion of premiums offered by the Southern Central Agricultural Society for silk and silk products. *Soil of the South*, I (March 1850), 238.
12. *Acts of the General Assembly* (1838), 222; *ibid.* (1840), 18-19.
13. *Compendium of the Sixth Census* (1840), 204.
14. *Federal Union* (Milledgeville), June 13, 1843 (quoted in U. B. Phillips, *A Documentary History . . .*, II, 291); *DeBow's Review*, I (May 1846), 434.

15. *Southern Cultivator*, IV (Jan. 1846), 5.
16. *Ibid.*, XIII (Oct. 1855), 151.
17. *Soil of the South*, I (Sept. 1851), 151.
18. *Federal Union*, June 12, 1843.
19. *Southern Cultivator*, III (Feb. 1845), 25.
20. *Ibid.*, XI (Oct. 1853), 302.
21. *Soil of the South*, I (July 1851), 99.
22. *Southern Cultivator*, XV (June 1857), 184.
23. *Soil of the South*, V (Dec. 1855), 360.
24. *Southern Cultivator*, VIII (Feb. 1850), 41.
25. *Ibid.*
26. *Ibid.*, III (Jan. 1845), 26; IX (June 1846), 88.
27. *Ibid.*, IV (June 1846), 88.
28. *American Farmer*, III (May 1821), 65; *Southern Cultivator*, II (Feb. 1844), 62; III (March 1845), 84.
29. *Southern Cultivator*, III (March 1845), 42, 143.
30. *Ibid.*, III (Feb. 1845), 40.
31. *Ibid.*, II (Feb. 1844), 84.
32. *Ibid.*, VII (Nov. 1849), 168; *Rome Courier*, April 24, 1851.
33. *Southern Cultivator*, IV (May 1846), 70.
34. *Soil of the South*, IV (March 1854), 99; *South Countryman*, I (Feb. 1859), 59.
35. *Southern Cultivator*, IV (May 1846), 70.
36. *Rome Courier*, Apr. 24, 1851.
37. *Southern Cultivator* XIV (Oct. 1856), 298; XV (April 1857), 120.
38. John Ruggles Cotting, *An Essay on the Soils and Available Manures of the State of Georgia* (Milledgeville, 1843), 51 (cited hereafter as *Essay on Soils*); *Southern Cultivator*, IX (Feb. 1853), 79; XVII (March 1859), 78.
39. *Southern Cultivator*, XII (Feb. 1854), 68 (quoted from the *Nashville Whig*).
40. *DeBow's Review*, VI (Oct.-Nov. 1848), 294.
41. *Southern Cultivator*, XIX (March 1861), 73.
42. John Stainback Wilson, *Woman's Home Book of Health* (Philadelphia, 1860), 127; *Southern Cultivator*, XVIII (August 1860), 295.
43. *Ibid.*, IV (Jan. 1848), 24.
44. J. S. Wilson, *Woman's Home Book of Health*, 127; *Southern Cultivator*, VIII (March 1849), 142; XVIII (Aug. 1860), 295. The standard price of Negro shoes was one dollar a pair. A single pair usually sufficed for a winter season. U. B. Phillips, *A Documentary History . . .*, I, 150-65.
45. *Southern Cultivator*, VII (March 1847), 92-93.
46. *Soil of the South*, IX (Feb. 1851), 28.
47. D. W. Lewis, *Transactions of the Southern Central Agricultural Society*, 136.
48. *Soil of the South*, II (Sept. 1852), 333.
49. *American Cotton Planter*, I (Nov. 1857), 370.
50. *Southern Cultivator*, XVII (April 1849), 52.
51. *Ibid.*, II (April 1844), 93.
52. *Ibid.*, VI (April 1848), 60-61.
53. S. P. Janes, *Geological Survey of Georgia. Bulletin No. 19* (Atlanta, 1909), 129-30; Mary A. Turner, Villa Rica, Ga., to Frances A. Williams, May 30, 1854 (in the Williams Collection of the Woman's College of Georgia Library.)
54. Herbert A. Kellar, *Solon Robinson . . .*, II, 454.
55. *Rome Courier*, Oct. 17, 1850.
56. Joseph Jones, *First Report to the Cotton Planters' Convention of Georgia . . .* (Augusta, 1860), 205.
57. *Southern Cultivator*, IV (March 1846), 36.
58. *Ibid.*, IV (June 1857), 184.
59. *Ibid.*, IV (Oct. 1856), 300; II (April 1844), 75.
60. *South Countryman*, I (March 1859), 73. State quarries in the vicinity of Rockmart and Cartersville were opened by Joseph G. Blance and Seaborn Jones around 1850. H. K. Shearer, *Geological Survey of Georgia. Bull. No. 34* (Atlanta, 1918), 1-2.
61. *Southern Cultivator*, XIX (May 1861), 150.

NOTES

62. *Ibid.*, X (May 1851), 77; XVI (May 1858), 142; XVI (June 1858), 176.
63. The estimated cost of fencing a 36-acre field in 1856 was $175. Post-oak fence posts brought thirty cents each. These high costs caused many to favor revision of the states' laws requiring fencing against free range livestock. *Southern Cultivator*, VIII (June 1850), 167; XIII (June 1855), 155.
64. *DeBow's Review*, VI (Oct.-Nov. 1848), 294.
65. *Acts of the General Assembly* (1847), 145.
66. J. R. Cotting, *Essay on Soils*, 51; *Southern Cultivator*, IX (Feb. 1853), 79; XVII (March 1859), 78.
67. *Southern Cultivator*, XII (Feb. 1854), 68.
68. Born in 1827 in Nova Scotia, William Gesner was the son of William Gesner, a Canadian scientist. Just before the outbreak of the Civil War he moved from Milledgeville to Montgomery where he was engaged in Confederate ordinance work and where he died in 1887. He was the author of several geological works on Alabama soils and minerals. *American Cotton Planter*, II (August 1858), 226; Thomas M. Owen, *History of Alabama*, 4 vols. (Chicago, 1921), III, 651.
69. *Southern Cultivator*, XVII (June 1859), 127; XVII (Sept. 1859), 264; *American Cotton Planter*, (August 1858), 256-62, 263.
70. *Southern Agriculturist*, II (March 1829), 98.
71. *Southern Cultivator*, I (April 1843), 41; II (Sept. 1844), 142.
72. *Ibid.*, III (Feb. 1845), 59.
73. *The Sorghum Handbook* (published in Cincinnati in 1888 for the Blymer Iron Company), 5-6; *Soil of the South*, V (Sept. 1855), 269; *Southern Cultivator*, XV (April 1857), 120.
74. *Ibid.*, XIII (April 1855), 130, 162; XV (April 1857), 120.
75. *Ibid.*, XIV (Oct. 1856), 298; XV (*April 1857*), 120.
76. *Ibid.*, XIII (April 1855), 130.
77. *Ibid.*, XIV (April 1856), 236, 345; *Federal Union*, Feb. 10, 1857.
78. *American Cotton Planter*, II (Feb. 1858), 56; *Southern Cultivator*, XVI (Feb. 1858), 86-87; XVII (July 1859), 198.
79. *Ibid.*, XIV (Oct. 1856), 313; XIV (Dec. 1856), 366. These early experiments have led many to err in assigning to Peters the credit for introducing sorghum into the South. Ex-Governor James H. Hammond of South Carolina reported a similar experiment a little earlier than that which occurred on the Peters farm. *The Atlanta Constitution*, Feb. 6, 1889; William L. Black, *A New Industry, or Raising the Angora Goat* . . . (Fort Worth, 1900), 60 (cited hereafter as *The Angora Goat*). For the prior claim of Hammond to the first sorghum experiment, see *Scientific American*, XIII (1856), 8.
80. *Southern Cultivator*, XIV (Oct. 1856), 298; (Nov. 1856), 332; (Dec. 1856), 366; *Georgia Citizen*, Oct. 11, 1856.
81. *Ibid.*, XIV (March 1856), 91; XV (Jan. 1857), 17, 109; Dennis Redmond, *Sorgho Sucre, or, Chinese Sugar Cane* . . . (Augusta, 1856), 1 *et passim*.
82. *South Countryman*, I (March 1859), 91; *The Countryman* (Turnwold, Georgia), VI (July 14, 1863), 6.
83. *The Sorghum Handbook*, 1.
84. *South Countryman*, (March 1859), 84.
85. "If a man lets his cattle get so thin that you may rivet a ten penny nail through their sides, he should not complain if they die when turned into a green pasture [of sorghum]," said one. *Southern Cultivator*, XVII (July 1859), 198.
86. *Ibid.*, XVI (Feb. 1858), 86-87; XXI (Nov.-Dec. 1863), 133.
87. *Ibid.*, XVI (Feb. 1858), 119.
88. *Ibid.*, XVI (April 1858), 112.
89. *Ibid.*, XVI (Feb. 1858), 87.
90. *The Sorghum Handbook*, 2.
91. *American Cotton Planter*, II (Feb. 1858), 56.
92. *The Sorghum Handbook*, 1; George W. Nichols, *The Story of the Great March* (New York, 1865), 58, 66; *Southern Cultivator*, XXI (May-

92. June 1863), 75. The popularity of sorghum as a human food was greatest in the Piedmont. In the Southern part of the state its use was associated with poverty and cheap diet.
93. Robert Nelson, "Chinese Sugar Millet," *American Cotton Planter*, II (Feb. 1858), 56.
94. *Ibid.*, I (Nov. 1857), 366; David Lewis Phares, *The Farmer's Book of Grasses and Other Forage Plants for the Southern United States* (Starkville, 1881), 19. Soy beans were introduced into the United States in 1804 but failed to become a permanent crop. The Perry expedition in 1854 brought from Japan two varieties of "soja beans" and the name "Japan pea" was used to designate the new product. This article underwent rediscovery about 1895 and came into prominence after World War I. William J. Morse and J. L. Carter, *Soybeans: Culture and Varieties. U. S. Dept. of Agric. Bull. No. 1520*, 2; *Yearbook of Agriculture* (1937), I, 55.
95. *Genesee Farmer*, VIII (Dec. 1847), 274; XVI (June 1855), 189; *Southern Cultivator*, VIII (May 1850), 68; XI (Jan. 1853), 40, 100, 111; XI (Oct. 1853), 314; *DeBow's Review*, X (Jan. 1853), 7.
96. J. William Jones thus quoted Lee in a speech heard by Douglas S. Freeman. Bell I. Wiley to James C. Bonner, May 26, 1942.
97. J. T. Henderson, *The Commonwealth of Georgia*, 62-63.
98. Dorothy Seay, "A Georgia Planter and His Plantations, 1837-1861," an unpublished master's thesis at the University of North Carolina (1937), p. 10 *et passim*.
99. Ralph B. Flanders, "Planters' Problems in Ante-Bellum Georgia," *Georgia Historical Quarterly*, XIV March 1930), 20.
100. *Southern Cultivator*, XIII (Oct. 1855), 305.
101. Appraisements of the Goods of William Terrell, Record Book "T," 6-20, Records of Hancock County, Court of Ordinary.
102. Emily Burke, *Reminiscences*, 220-31.
103. H. A. Kellar, *Solon Robinson*, 481.
104. *Southern Cultivator*, XVIII (Sept. 1860), 297.
105. *Ibid.*, XI (Sept. 1853), 294.
106. *South Countryman*, I (Feb. 1859), 38.
107. *Rome Courier*, May 27, 1852; *Soil of the South*, IV (July 1854), 199; *American Cotton Planter . . .*, II April 1858), 181-82; *Southern Cultivator*, XVIII (July 1860), 211.
108. *Ibid.*, XII (May-June, 1863), 77.
109. *Southern Agriculturist*, I (Aug. 1828), 357.
110. *Ibid.*, I (Oct. 1828), 433.
111. *Soil of the South*, II (July 1852), 290. Suggestions were made in 1860 to include on the premium list of the Georgia State Agricultural Society articles made by Negro artisans, since skilled Negro craftsmen were found in almost every plantation community. *Southern Cultivator*, XVIII (April 1860), 250.
112. *Ibid.*, VI (Oct. 1848), 169; H. A. Kellar, *Solon Robinson*, II, 473.
113. *Southern Cultivator*, VIII (July 1850), 102-103.
114. The Eighth Census of the United States, Schedule IV.
115. *Southern Cultivator*, XVII (Sept. 1859), 284.
116. *Soil of the South*, I (Jan. 1851), 7.
117. Robert P. Brooks, "Mark Anthony Cooper," in *Dictionary of American Biography*, IV, 407-08.
118. Lucy Josephine Cunyus, *History of Bartow County* (Cartersville, 1933), 279.
119. *South Countryman*, I (Feb. 1859), 43-44 (quoted from the *Cartersville Express*).
120. *Ibid.*, I (Feb. 1859), 44.
121. H. A. Kellar, *Solon Robinson*, II, 469; Nellie Peters Black, *Richard Peters, His Ancestors and Descendants* (Atlanta, 1904), 72; *Manufactures of the United States* (1860), 62, 82.
122. *Rome Courier*, July 8, 1852.
123. T. B. Thorpe, "Cotton and its Cultivation," *Harper's New Monthly Magazine*, III (March 1854), 463.

NOTES 217

CHAPTER VII

1. *Southern Agriculture*, III (March 1845), 122; *Southern Cultivator*, III (March 1845), 99, 100. The barshare was the point of a mold board plow, usually welded to the landside, which cut the furrow slices at the side and bottom. Its reputation for shallow plowing is indicated by the following statement: "Absence from home . . . compelled me to depend . . . on annual 'cropers,' who were accustomed to skim over the ground with the 'bar share' plow." Illinois Agricultural Society, *Transactions*, III (1957-1958), 408.
2. *Southern Cultivator*, II (June 1844), 98; V (Jan. 1747), 10; *American Agriculturist*, II (Feb. 1843), 49.
3. *Ibid.*, III (May 1844), 247; *Southern Cultivator*, II (June 1844), 98; IV (Feb. 1846), 39; V (March 1847), 90; XV (March 1858), 92.
4. *Albany Cultivator*, IX (June 1842), 93.
5. J. R. Cotting, *Essay on Soils*, 51.
6. *The Horticulturist*, VI (April 1851), 156.
7. *Albany Cultivator*, IV (May 1847), 156.
8. H. A. Kellar, *Solon Robinson*, II, 478.
9. *Ibid.*, 476.
10. *Southern Cultivator*, IV (April 1846), 59.
11. *Ibid.*, II (June 1844), 98.
12. *Ibid.*, X (Feb. 1852), 36.
13. *Ibid.*, VI (April 1848), 52.
14. *Ibid.*, VI (May 1848), 70.
15. *Ibid.*, VI (June 1848), 89.
16. *Ibid.*, XVI (July 1858), 208.
17. A. F. Hopkins to *DeBow's Review*, quoted in *Southern Cultivator*, XIX (Jan. 1861), 14.
18. David A. Wells, *Yearbook of Agriculture*, 1855 and 1856 (Philadelphia, 1856), 388; *Southern Cultivator*, XIII (Sept. 1856), 273, 274.
19. *Ibid.*, XIII (Dec. 1856), 354.
20. Garland D. Harmon, "Cotton Plow and Scraper," *Southern Cultivator*, XIII (Sept. 1856), 274.
21. L. C. Gray, *op. cit.*, II, 794-96; *Southern Cultivator*, XVIII (Feb. 1860), 44; XIX (May 1861), 298.
22. *The Rural Southerner*, IV (April 1871), 97.
23. *Farmer and Planter*, XI (Dec. 1860), 306. Southerners were far behind the rest of the country in making any kind of mechanical inventions, however. A study of the Patent Office records fails to reveal any noteworthy invention by Georgians.
24. *Southern Cultivator*, XVIII (Feb. 1860), 54.
25. *Ibid.*, III (March 1845), 100.
26. *American Farmer*, I (Feb. 1820), 358; *Farmer's Register*, I (Dec. 1834), 680.
27. *Southern Agriculturist*, III (March 1830), 123.
28. *Southern Cultivator*, XVII (Aug. 1859), 262.
29. *Ibid.*, II (June 1844), 98-99; L. C. Gray, *op. cit.*, II, 801.
30. *Farmer's Register*, X (Aug. 1842), 1-2; *Southern Cultivator*, II (Jan. 1844), 1-2.
31. *Ibid.*, III (June 1845), 87; IX (May 1851), 68-69; *Farmer's Register*, X (Aug. 1842); 390-91.
32. *American Cotton Planter . . .*, I (July 1857), 199; (Aug. 1857), 234; *Soil of the South*, I (Oct. 1851), 146; *Southern Farm and Home*, I (Nov. 1869), 5; *Southern Cultivator*, XIII (Oct. 1856), 304.
33. *Ibid.*, II (June 1844), 98-99.
34. *Ibid.*, XIII (Oct. 1856), 304.
35. *Ibid.*, XVIII (Nov. 1860), 414, quoted from the Sandersville *Central Georgian*.
36. *Farmer's Register*, X (Aug. 1842), 390; *Southern Cultivator*, II (Jan. 1844), 1-2.
37. *American Farmer*, X (Aug. 1842), 390-91.
38. *Ibid.*, I (Feb. 1820), 358; X (Aug. 1842), 390; *American Cotton Planter . . .*, I (July 1857), 198.
39. *Ibid.*, I (Aug. 1857), 240; II (Jan. 1858), 18; *Farmer's Register*, I (July 1834), 334.
40. Edmund M. Pendleton, *Textbook of Scientific Agriculture* (New York, 1876), 140; *American Cotton Planter . . .*, I (Aug. 1857), 240; *Southern Cultivator*, XIV (July 1856), 215.

41. H. A. Kellar, *Solon Robinson*, II, 478.
42. *Southern Cultivator*, XV (Feb. 1857), 53.
43. *Federal Union*, April 23, 1850, quoted in U. B. Phillips, *A Documentary History*, II, 132.
44. *Southern Cultivator*, VI (Feb. 1848), 24; XVI (Aug. 1857), 250.
45. Ditches and guide rows could be laid off at the average price of two dollars an acre, with the planter providing two assistants for the surveyor and board for three men and a horse. *Southern Cultivator*, XIV (Oct. 1857), 298.
46. Andrew B. Booth, ed., *Records of Louisiana Confederate Soldiers and Louisiana Confederate Commands*, 3 vols. (New Orleans, 1920), III, Book I, 192; *Southern Cultivator*, XI (March 1853), 78-79; *Soil of the South*, IV (July 1854), 199; Records of Polk County, Deed Book "C" (Clerk of Superior Court), 280-82.
47. Records of Floyd County, Deed Book "L" (Clerk of Superior Court), 561-62; Annual Returns, 1846-1847 (Court of Ordinary), 181.
48. *Ibid.*, Deed Record "J," 59.
49. *Rome Courier*, Nov. 7, 1851, Feb. 26, 1852, May 27, 1852, July 8, 1852; *Southern Cultivator*, IX (July 1852), 201-02.
50. *Ibid.*, XIII (July 1856), 215-236; (Sept. 1856), 272; (May 1859), 142; XVIII (Nov. 1860), 343; *Soil of the South*, IV (July 1854), 199.
51. *Southern Cultivator*, XVII (May 1859), 142.
52. *Soil of the South*, (March 1854), 73; (July 1854), 199.
53. *Southern Cultivator*, XI (Jan. 1853), 9.
54. *Ibid.*, XIII (Sept. 1856), 273.
55. *Ibid.*, XVI (April 1858), 273, 274.
56. *Ibid.*, XV (July 1857), 224; XVII (July 1859), 197; *American Cotton Planter*, II (April 1858), 81-82.
57. *Ibid.*, II (Dec. 1859), 371.
58. *Southern Cultivator*, XVI (April 1858), 148; (July 1858), 210; XVII (July 1859), 210.
59. *Ibid.*, XVII (Aug. 1859), 235.
60. *American Cotton Planter*, I (June 1857), 210.
61. *Ibid.*, II (July 1858), 209; *Southern Cultivator*, XV (March 1859), 83.
62. *Ibid.*, XIV (Sept. 1856), 273; *Soil of the South*, II (Sept. 1852), 329.
63. *Southern Cultivator*, XV (March 1857), 83; (Aug. 1857), 250.
64. Franklin L. Riley, ed., "Diary of a Mississippi Planter," Jan. 1, 1840 to April, 1863, *Publications of the Mississippi Historical Society*, X, 460.
65. *Ibid.*, X, 460; *American Cotton Planter*, II (April 1858), 181-82; *Southern Cultivator*, XVI (April 1858), 144.
66. *American Cotton Planter*, II (Jan. 1854), 14, 213.
67. *Ibid.*, II (Sept. 1854), 280; (Nov. 1854), 346.
68. *Southern Cultivator*, VIII (Dec. 1849), 181.
69. *Ibid.*, XIV (April 1856), III, 113; (Oct. 1857), 300; *American Cotton Planter*, I (April 1857), 328.
70. *Southern Cultivator*, XIV (April 1856), 111.
71. *Ibid.*, XIX (April 1861), 119.
72. *Ibid.*, XVI (April 1858), 144; XVIII (May 1860), 151; (Nov. 1860), 85.
73. *Ibid.*, XVIII (Feb. 1860), 78; (July 1860), 211, 214; *American Cotton Planter*, I (Aug. 1857), 238; II (Oct. 1858), 310.
74. *Ibid.*, I (Aug. 1857), 238.
75. *Southern Cultivator*, VIII (Sept. 1850), 135; XIX (May 1861), 151; VII (May 1849), 75.
76. *Ibid.*, VIII (April 1861), 119.
77. *Ibid.*, XIX (Sept. 1861), 119; A. B. Booth, *Records of Louisiana Confederate Soldiers . . .*, III, 192.
78. *Ibid.*, *Southern Cultivator*, XXI (May-June 1863), 77.
79. *Ibid.*, XXI (Nov.-Dec. 1863), 128; XXII (March 1864), 51.
80. *Ibid.*, XXI (Nov.-Dec. 1863), 128; XXII (March 1864), 51.
81. *Ibid.*, XXV (March 1867), 68. For a more complete discussion of Garland D. Harmon, see James C. Bonner, "The Plantation Overseer and Southern Nationalism as Revealed in the Career of Garland D. Harmon," *Agricultural History*, XIX (Jan. 1945), 1-11).

NOTES 219

CHAPTER VIII

1. Prince's *Digest*, 585.
2. *American Farmer*, II (March 1821), 412.
3. *Ibid.*, VII (Aug. 1825), 185-86; E. M. Coulter, *Thomas Spalding of Sapelo*, 47-49.
4. *American Farmer*, VII (April 1825), 41; (Aug. 1825), 42; *Southern Agriculturist*, I (Feb. 1828), 57.
5. *Ibid.*, I (July 1828), 308; *Prince's Digest*, 974.
6. R. T. Nesbitt, *Georgia: Her Resources and Possibilities* (Atlanta, 1895), 100 (cited hereafter as *Georgia*); E. Merton Coulter, *A Short History of Georgia* (Chapel Hill, 1933), 264; Prince's *Digest*, 884. For a discussion of the Georgia Board of Public Works see Fletcher M. Green, "Georgia's Board of Public Works," *Georgia Historical Quarterly*, XXII (June 1938), 117-37.
7. *Acts of the General Assembly* (1830), 20, 197, 206. The office of the State Geologist was recreated in 1874 and again abolished, but the work was finally resumed toward the end of the century. Nesbitt, *Georgia*, 101.
8. *Acts of the General Assembly* (1838), 222.
9. *Compendium of the Sixth Census* (1840), 204; *Acts of the General Assembly* (1840), Appendix, 19; Executive Minutes, Appropriations and Disbursements, 1835-1840, in Georgia Department of Archives and History.
10. *Southern Cultivator*, VII (Jan. 1849), 11.
11. Roland M. Harper, "Development of Agriculture in Upper Georgia from 1850 to 1880," *Georgia Historical Quarterly*, VI (March 1922), 14, 15; Mrs. Elizabeth Smith, A History of Hancock County (a manuscript in possession of Mrs. Sara Carnes, of Atlanta, Ga.), 23 *et passim*.
12. Smith, A History of Hancock County, *loc. cit.*, 1 *et passim*.
13. The organization of the Hancock Planters Club was followed by clubs in Greene, Morgan, Monroe, Wilkes, and Oglethorpe counties. By 1846 clubs existed in Putnam, Warren, Habersham, Early, Burke, Clarke, Jefferson, DeKalb, Walton, Liberty, Harris, Baker, and Muscogee. In general it was not until the 1850's that agricultural societies became popular in the newer counties of the western and northern areas of the State. See *Soil of the South*, I (Sept. 1851), 139; II (Jan. 1852), 3; (April 1852), 251; (July 1851), 104; (May 1851), 260; *Southern Cultivator*, III (June 1845), 93; (Sept. 1845), 142; IV (Nov. 1846), 191; XVIII (July 1860), 214; *South Countryman*, I (Jan. 1859), 26.
14. *Southern Cultivator*, XVIII (July 1860), 214.
15. *Ibid.*, III (June 1845), 93; XVII (Dec. 1860), 341; Records of the Hancock Planters Club, Minutes, 1837, in custody of the Clerk of Superior Court, Sparta, Ga.; Elisha Hunter to Tuttle H. Audas, Nov. 25, 1856, *loc. cit.*
16. *Southern Cultivator*, III (March 1845), 77; (June 1845), 93; VII (April 1849), 60.
17. *Ibid.*, I (Dec. 1843), 206; III (June 1845), 94, 110; II (April 1852), 25; VIII (July 1858), 104.
18. Records of the Hancock Planters Club, Minutes, 1837.
19. *Ibid.*, Records of the Executive Department of Georgia, County Officers, 1836-61; Productions of Agriculture in Hancock County in 1850 (Schedule IV, Census of 1850), a manuscript in the Duke University Library.
20. Productions of Agriculture in Hancock County in 1860 (Schedule IV, Census of 1860, *loc. cit.*); Slave Inhabitants in Hancock County in 1860 (Schedule II, Census of 1860), a manuscript in custody of the U. S. Dept. Of Commerce, Washington, D. C.; *Southern Cultivator*, XVIII (Dec. 1860), 388.
21. Ralph B. Flanders, *Plantation Slavery in Georgia* (Chapel Hill, 1933), 142, 272; Records of Hancock County, Will Book "I," Vol. 1, 454.

22. *Soil of the South*, IV (May 1854), 165.
23. Records of the Hancock Planters Club, Constitution (1837); Treasurer's Report (June 5, 1847); Minutes (1844-1845).
24. *Ibid.*, Minutes (1841-1844); *Southern Cultivator*, III (March 1845), 42; VI (Jan. 1846), 5.
25. *Southern Cultivator*, II (Feb. 1844), 22.
26. *Ibid.*, XVII (Oct. 1859), 261, 262; (Dec. 1859), 367; *South Countryman*, I (Jan. 1859), 26.
27. *Southern Cultivator*, III (Dec. 1845), 180-81.
28. *Ibid.*, I (March 1843), 62; III (Dec. 1845), 180.
29. *Ibid.*, I (Sept. 1843), 141, 151; XVIII (Dec. 1860); 391; *South Countryman*, I (Jan. 1859), 26.
30. *Soil of the South*, I (Jan. 1851), 3, 15-16.
31. *Ibid.*, I (Sept. 1851), 139.
32. *Rome Courier*, July 8, 1852, Sept. 16, 1862; Journal of Eugene Le Hardy, 1851-1853 (a manuscript in the De Renne Collection, University of Georgia Library), Feb. 1851, p. 28.
33. *Southern Cultivator*, II (July 1844), 120; III (May 1845), 77; (Dec. 1845), 179.
34. *Ibid.*, V (Jan. 1847), 113, 184.
35. D. W. Lewis, *Transactions*, i-xi, 257.
36. *Ibid.*, ix-x, xiv, 277; Rebecca Latimer Felton, *Country Life in Georgia in the Days of My Youth* (Atlanta, 1919), 69.
37. R. T. Nesbitt, *Georgia*, 267; D. W. Lewis, *Transactions*, iii, *South Countryman*, I (March 1859), 70; *Soil of the South*, V (Oct. 1855), 289; VI (April 1856), 100.
39. *Southern Cultivator*, VI (April 1849), 60; John C. Butler, *Historical Record of Macon and Central Georgia* (Macon, 1879), 189 (cited hereafter as *Macon*); *Acts of the General Assembly* (1849-1850), 203.
40. *Southern Cultivator*, XI (July 1853), 209; (Aug. 1853), 243, 273; (Dec. 1853), 372.
41. *Ibid.*, XVI (Sept. 1858), 286; D. W. Lewis, *Transactions*, xiv.
42. *Soil of the South*, II (Oct. 1853), 341; *Southern Cultivator*, X (Aug. 1852), 248; *Georgia Citizen* (Macon), June 26, 1852. The Georgia Silk and Agricultural Society was founded at Columbus in 1839 during the revival of interest in silk production. It was short-lived. *Georgia Jeffersonian* (West Point), March 18, 1840.
43. A comparison of the agricultural census of 1850 with that of 1860 indicates that large planters were moving more definitely toward the single crop system than were the small farmers.
44. *Southern Cultivator*, XVIII (April 1860), 132. Throughout the fifties agricultural leadership was divided and it displayed the same kind of schismatic symptoms which characterized Southern political life of this decade. In addition to the Cotton Planters Convention at Macon, the Agricultural Convention of the Slaveholding States was formed at Montgomery in 1853 at which time Robert Toombs and William L. Yancey played leading roles and of which ex-Governor George R. Gilmer was elected president. *Soil of the South*, I (Dec. 1851), 178; III (July 1853), 564, 586, 589.
45. *Southern Cultivator*, XVIII (April 1860), 240; XIX (May 1861), 154.
46. *Ibid.*, XV (April 1857), 146; XVIII (Aug. 1860), 248; (Nov. 1860), 352-54.
47. *Soil of the South*, I (Jan. 1851), 10; Albert L. Demaree, *The American Agricultural Press, 1819-1860* (New York, 1941), 372.
48. *Soil of the South*, III (April 1853), 497; *American Cotton Planter*, I (Jan. 1857), 1.
49. *South Countryman*, I (Jan. 1859), 2.
50. *Georgia Jeffersonian*, March 18, 1840; *Carolina Planter*, I (May 1840), 129.
51. *The American Cotton Planter*, II (Dec. 1858), 377; *Literary Companion* (Newnan), Jan. 23, 1861.
52. A. L. Demaree, *American Agricultural Press*, 375; *Southern Cultivator*, II (Nov. 1844), 183; IV (May 1846), 72.
53. *Southern Field and Fireside* (Augusta), II (July 1860), 62, 158;

Southern Cultivator XVI (May 1858), 138; XVII (March 1859), 83; A. L. Demaree, American Agricultural Press, 373.
54. Southern Cultivator, XV (April 1857), 146; XVIII (Nov. 1860), 352-53.
55. Emily G. Ravenel, Record of Charles Wallace Howard, an unpublished manuscript in private possession, 1 et passim; Lucy J. Cunyus, A History of Bartow County, 41 et passim; South Countryman, I (Jan. 1859), 1; Allen P. Tankersley, College Life at Old Oglethorpe (Athens, 1951), 5, 7, 24.
56. Southern Cultivator, III (March 1845), 77; XI (Dec. 1853), 372, 374.
57. Southern Field and Fireside, I (Feb. 1860), 319.
58. D. W. Lewis, Transactions, 163-76.
59. "Educational Convention," a broadside published at Macon, April 21, 1851.
60. D. W. Lewis, Transactions, 176; Southern Cultivator, XVII (Aug. 1859), 226-30.
61. Soil of the South, V (Dec. 1855), 373.
62. Southern Cultivator, V (Sept. 1847), 141.
63. Soil of the South, IV (Feb. 1854), 41; Joseph Henry Lumpkin, An Address Delivered Before the South Carolina Institute, . . . Nov. 19, 1850 (a pamphlet published at Charleston in 1851), 32-38.
64. Southern Cultivator, III (Sept. 1845), 141.
65. South Countryman, I (Jan. 1859), 13; American Cotton Planter, I (Dec. 1857), 374; Southern Cultivator, XVII (Aug. 1859), 240; Soil of the South, III (Nov. 1855), 325.
66. American Cotton Planter, I (Dec. 1857), 375.
67. Southern Cultivator, XVII (Aug. 1859), 240.
68. E. Merton Coulter, "A Georgia Educational Movement During the Eighteen Hundred Fifties," Georgia Historical Quarterly, IX (March 1925), 30.
69. Genesee Farmer, XV (Sept. 1854), 289, 299; Southern Cultivator, XIII (Aug. 1855), 250. John P. Norton's gift of $5,000 to Yale preceded Terrell's gift to Franklin College. Only two donations of a thousand dollars each had been made to Franklin College by private persons before 1854. Augustus L. Hull, Annals of Athens, Georgia, 1801-1901. (Athens, 1906), 28; Ibid., A Historical Sketch of the University of Georgia (Athens, 1894).

CHAPTER IX

1. American Farmer, III (May 1821), 52; Southern Agriculturist, II (June 1829), 274-275; III (Dec. 1830), 624; A. L. Demaree, American Agricultural Press, 157.
2. Southern Agriculturist, III (June 1830), 303; Soil of the South, I (Dec. 1851), 185; James Thomas, "Winter Grasses for the South," Southern Cultivator, IX (May 1851), 76; Charles W. Howard, A Manual of the Cultivation of the Grasses and Forage Plants at the South (Atlanta, 1881), 17. Cited hereafter as Manual of Grasses.
3. Southern Cultivator, IX (May 1851), 72; VII (July 1849), 109; XVIII (Aug. 1860), 245; South Countryman, I (Feb. 1859), 57; Charles Lanman, Letters from the Allegheny Mountains (New York, 1849), 53. Cited hereafter as Letters.
4. John Lord Hayes, Sheep Husbandry in the South (Boston, 1878), 100 (cited hereafter as Sheep Husbandry); E. M. Pendleton, Textbook of Agriculture, 397; C. W. Howard, A Manual of Grasses, 24; Southern Cultivator, XVI (April 1858), 130. Timothy grass is said to have received its name from the fact that it was brought to the South by Timothy Hanson. It was also known as catstail grass. John Nicholson, The Farmer's Assistant (Richmond), 385; American Farmer, I (Nov.

1820), 390; David L. Phares, *The Farmer's Book of Grasses and Other Forage Plants for the Southern United States* (Starkville, 1881), 35.

5. C. W. Howard, *Manual of Grasses*, 16, 17; D. L. Phares, *The Farmer's Book of Grasses*, 47; *Southern Cultivator*, X (Aug. 1852), 252; XVIII (July 1860), 219; *Southern Agriculturist*, III (June 1830), 303; *South Countryman*, I (Jan. 1859), 27.

6. *Soil of the South*, III (Aug. 1853), 625; XIII (May 1855), 142-43; *American Cotton Planter*, II (Jan. 1858), 34; *Report of the Commissioner of Patents for the Year 1853. Agriculture*, 212.

7. *Soil of the South*, IV (Oct. 1845), 295; *Southern Cultivator*, XII (Nov. 1854), 337; XIII (May 1855), 142-43.

8. C. W. Howard, *Manual of Grasses*, 18-19; Liberty H. Bailey, *The Standard Cyclopedia of Horticulture*, 3 vols. (New York, 1930), I, 578.

9. *Southern Cultivator*, XII (Nov. 1854), 337. It is possible that Iverson may have been sincere in his claims. One Professor Torrey of New York identified this grass as a distinct variety, naming it *Ceratochloa Breviaristata* or "Short Awn Horn Grass." Iverson claimed it to be a Russian grass which had become native to the Pacific Coast. *Southern Cultivator*, XIII (March 1855), 70; (July 1855), 225.

10. *Southern Cultivator*, VIII (March 1849), 40.

11. *Genesee Farmer*, IX (Sept. 1848), 214.

12. *Southern Cultivator*, III (Jan. 1845), 13. According to Spalding, Bermuda was first brought to Savannah by Governor Henry Ellis in 1757.

13. C. W. Howard, *Manual of Grasses*, 17; *Southern Cultivator*, XI (Jan. 1855), 238; *South Countryman*, I (Jan. 1859), 27.

14. *Southern Cultivator*, III (March 1845), 57; (April 1845), 107. "One undigested joint in a cow's stomach when dropped is sufficient to ruin the best plantation in Georgia," said a Middle Georgia farmer. *Ibid.*, VIII (Jan. 1849), 21.

15. *Southern Cultivator*, III (March 1845), 76, 91; VI (Nov. 1848), 166; VII (Jan. 1849), 21.

16. *Ibid.*, III (May 1845), 76; (Sept. 1845), 138-39; (Oct. 1845), 153; XVII (Nov. 1860), 332; E. M. Pendleton, *Textbook of Grasses*, 397; John L. Hayes, *Sheep Husbandry in the South*, (Boston, 1878), 100.

17. *Southern Cultivator*, VI (Dec. 1848), 180.

18. *Ibid.*, XVIII (Sept. 1860), 265-66.

19. C. W. Howard, *Manual of Grasses*, 5, 19, 21, 35; *Southern Cultivator*, VII (Feb. 1849) 42; XIV (Jan. 1856), 25; XVIII (June 1859), 162; (April 1860), 122; (Oct. 1860), 229; Ansten Anstensen, *The Proverb in Isben*, 3 *et passim*.

20. *South Countryman*, I (Jan. 1859), 2; C. W. Howard, *Manual of Grasses*, 7.

21. *Southern Cultivator*, XVI (Feb. 1858), 275; C. W. Howard, *Manual of Grasses*, 10; *South Countryman*, I (Jan. 1859), 27.

22. C. W. Howard, *Manual of Grasses*, 8; J. L. Hayes, *Sheep Husbandry*, 20; *Southern Cultivator*, XVIII (Nov. 1860), 330-31.

23. *Ibid.*, IV (July 1846), 108; XIX (July 1861), 216; *Soil of the South*, XI (Jan. 1853), 397; *South Countryman*, I (Jan. 1859), 9; N. P. Black, *Richmond Peters*, 53; *U. S. Department of Agriculture Yearbook for 1937*, 1124.

24. N. P. Black, *Richard Peters*, 36.

25. C. W. Howard, *Manual of Grasses*, 10.

26. *The Banner Herald* (Athens), June 5, 1931; Records of Peters Land Co., Deeds, Peters' Account Book; N. P. Black, *Richard Peters*, 16, 17, 18.

27. *Ibid.*, 35, 53, H. A. Kellar, *Solon Robinson*, II, 469; *Southern Cultivator*, XIV (Oct. 1856), 322.

28. *Ibid.*, XIX (July 1861), 216; C. W. Howard, *Manual of Grasses*, 135; N. P. Black, *Richard Peters*, 35.

29. *Ibid.*, 44.

30. Wash J. Houston to N. P. Black, Dec. 23, 1901, in the Peters Collec-

tion of the University of Georgia; Records of Peters Land Co., Farm Record Book, 1841. Apparently little effort was made to improve the native cattle which roamed the vast area of open range in southern Georgia. Here the cattle industry was conducted on much the same principles as those which characterized the Colonial era and those in the West a century later. The occupation of "cow hunter" was common in this area. It is significant to note that a Confederate company organized in Irwin County in 1861 chose as their official name the "Irwin Cow Boys." This is one of the earliest uses of the term "cow boy." Chas. E. Jones, *Georgia in the War, 1861-1865*, (Atlanta, 1909), 16.
31. *Richmond Enquirer*, July 2, 1833, quoted in *Farmer's Register*, I (May 1834), 170.
32. *Southern Cultivator*, III (Aug. 1845), 121.
33. *Ibid.*, V (Jan. 1847), 7; XX (July-Aug. 1862), 142; N. P. Black, *Richard Peters*, 38.
34. *Ibid.*, 38, 52; *Southern Cultivator*, XIX (July 1861), 216.
35. George W. Williams, *A Sketch of Nacoochee, Georgia and Its Surroundings* (n.d., a pamphlet in the DeRenne Collection, University of Georgia Library), 50; Charles Lanman, *Letters*, 53; *Southern Cultivator*, VI (May 1849), 70; *Rural Carolinian*, I (Oct. 1869), 453.
36. *South Countryman*, I (Feb. 1859), 57; *Southern Cultivator* VII (May 1849), 76; XVIII (Aug. 1860), 245; D. W. Lewis, *Transactions*, 330-31.
37. *Southern Cultivator*, XX (July-Aug. 1862), 142.
38. See Schedule IV, Census of 1860, Productions of Agriculture for various counties, *loc. cit.* In 1856 William Terrell possessed 26 mules, 20 oxen, and only nine horses. Records of Hancock County, Record Book "T," Court of Ordinary, 6-20.
39. U. B. Phillips, *Life and Labor*, 134-35; *Genesee Farmer*, IX (Feb. 1848), 37.
40. *Seventh Census of the United States, Agriculture*, 378; *Eighth Census of the United States, Agriculture*, 26; *De Bow's Review*, VI (Oct.-Nov. 1848), 294; *Southern Cultivator*, XI (Nov. 1853), 34; XIII (Jan. 1855), 11; *Soil of the South*, V (Oct. 1855), 293.
41. *Soil of the South*, V (Sept. 1855), 261; *Southern Cultivator*, VII (April 1849), 50; N. P. Black, *Richard Peters*, 43.
42. *Southern Cultivator*, XVII (Sept. 1859), 270, 271; XX (July-Aug. 1862), 142.
43. *Soil of the South*, V (Oct. 1855), 293; *Southern Cultivator*, XV (Dec. 1857), 363.
44. *South Countryman*, I (Feb. 1859), 58; *Southern Cultivator*, XI (Jan. 1853), 17; XVIII (Aug. 1860), 245.
45. J. L. Hayes, *Sheep Husbandry*, 3; Henry S. Randall, *Sheep Husbandry in the South* (Philadelphia, 1848), 7 (cited hereafter as *Sheep Husbandry*); *Compendium of the Sixth Census, 1840*, 198. The sheep census of Georgia in 1840 shows a large number of these animals in the older counties of the Piedmont Cotton Belt as well as in the newer counties of the extreme western part of the state; namely, Carroll, Coweta, Troup, Talbot, and Meriwether counties.
46. The principal breeds of improved sheep in the northern United States at this period were the Saxon merinos, the Bakewell, Southdown, Cotswold, Lincoln, and Cheviot, all of which came from England. In addition to these were the Spanish merino and the "African broad tail." H. S. Randall, *Sheep Husbandry*, 7, 8.
47. *Ibid.*, 7; *Southern Cultivator*, III (March 1845), 73; (August 1845), 156.
48. Quoted in *Southern Cultivator*, III (March 1845), 73.
49. *Ibid.*, III (Jan. 1845), 25, IV (Jan. 1846), 4; XL (Sept. 1853), 279; *South Countryman*, I (Jan 1859), 19.
50. J. L. Hayes, *Sheep Husbandry*, 84.
51. H. S. Randall, *Sheep Husbandry*, 64; *Southern Cultivator*, IV (Jan.

1846), 9. A single day's entry in Peters' Sheep Registry Book showed ten sheep killed by dogs and eleven dead of unknown causes. Records of Peters' Land Co.
52. *Genesee Farmer,* IX (Feb. 1848), 37.
53. *Southern Cultivator,* XX (Nov.-Dec. 1862), 200; XXI (Jan.-Feb. 1863), 12; XXI (Nov.-Dec. 1863), 136; *Acts of the General Assembly* (1862), p. 111.
54. *Genesee Farmer,* IX (Sept. 1848), 216.
55. N. P. Black, *Richard Peters,* 41; *South Countryman,* I (Jan. 1859), 19; *Southern Cultivator,* XIX (July 1861), 216.
56. Richard Peters' Sheep Registry Book, *loc. cit.; South Countryman,* I (Jan. 1851), 19; N. P. Black, *Richard Peters,* 41; Nellie Peters Black scrapbook in the Peters Collection, *loc. cit.*
57. J. L. Hayes, *Sheep Husbandry,* 29; Richard Peters to John L. Hayes, *ibid.,* 99, 101.
58. H. S. Randall, *Sheep Husbandry,* 155; *Southern Cultivator,* XIX (July 1861), 216.
59. N. P. Black, *Richard Peters,* 41; J. L. Hayes, *Sheep Husbandry,* 32; N. P. Black Scrapbook, *loc. cit.;* William L. Black, *A New Industry, or Raising the Angora Goat, and Mohair for Profit* (Fort Worth, 1900), 52, (cited hereafter as *The Angora Goat*); *Report of the Commissioner of Patents, Agriculture* (1853), 20; *American Farmer,* X (May 1828), 83.
60. W. L. Black, *The Angora Goat,* 192, 313, 314; N. P. Black, *Richard Peters,* 42; Nellie Peters Black Scrapbook, *loc. cit.*
61. *New Orleans Picayune,* March 19, 1859, quoted in *Southern Cultivator,* XVIII (April 1859), 187; N. P. Black, *Richard Peters,* 42; W. L. Black, *The Angora Goat,* 192.
62. *Ibid.,* 177; J. L. Hayes, *Sheep Husbandry,* 51; N. P. Black Scrapbook, *loc. cit.*
63. N. P. Black, *Richard Peters,* 41.
64. W. L. Black, *The Angora Goat,* 177, 250; J. L. Hayes, *Sheep Husbandry,* 51, 96; N. P. Black Scrapbook, *loc. cit.* One of Peters' bucks, "Mahomet," was described by Henry W. Grady as follows: "A million debauches have writ their shocking story across his narrow brow. A million wrinkles have furrowed his sensual jaws. . . . A dull fatalism burns in his introverted eyes, where once the light of Asia flowed, and Dejection sits listless where Desire once hung its flaunting banners. A wisp of ancient hair . . . straggles above his brow as funereal moss above a tomb." *Atlanta Constitution,* Oct. 12, 1884.
65. W. L. Black, *The Angora Goat,* 218; J. L. Hayes, *Sheep Husbandry,* 199; N. P. Black Scrapbook, *loc. cit.*
66. W. L. Black, *The Angora Goat,* 92, 110, 313; Richard Peters to John L. Hayes, Jan. 1, 1878, in N. P. Black Scrapbook; *Atlanta Constitution,* Feb. 6, 1889. William Landrum went to California from Georgia in 1850 and was among the first to recognize the agricultural possibilities of the San Joaquin Valley where he began farming in 1852. After the Civil War, he established herds of Angoras in various Western states. W. L. Black, *The Angora Goat,* 79, 92.
67. *Ibid.,* 52, 93, 96, 344; N. P. Black Scrapbook, *loc. cit.* By 1862 the Angora goat had been raised successfully in thirteen states east of the Mississippi. Only the New England states had shown no interest in the industry.
68. W. L. Black, *The Angora Goat,* 79, 98, 303, 446; Ralph Peters Black to James C. Bonner, Aug. 19, 1941; *Yearbook of the U. S. Department of Agriculture, 1898, Division of Statistics,* 425-426.
69. *Southern Cultivator,* XIX (July 1861), 216. For a more detailed description of Peters' Angora experiments, see James C. Bonner, "The Angora Goat: A Footnote in Southern Agricultural History," *Agricultural History,* XXI (Jan. 1947), 42-46.
70. W. L. Black, *The Angora Goat,* 109, 114, 203; J. L. Hayes, *Sheep Husbandry,* 66, 67.
71. *Farmer's Register,* III (Dec. 1835),

495; *Southern Cultivator*, II (Feb. 1844), 18.
72. Percy W. Bidwell and John I. Falconer, *History of Agriculture in the Northern United States, 1820 to 1860*. (Washington, 1925), 230, 441; *Southern Cultivator*, I (May 1843), 71; (August 1843), 121.
73. *Albany Cultivator*, I (Jan. 1844), 14.
74. *Ibid.*, 23; *American Cotton Planter . . .*, I (Aug. 1857), 263; *Southern Cultivator*, I (Dec. 1843), 190.
75. *Ibid.*, III (Aug. 1845), 121.
76. *Ibid.*, VII (Jan. 1850), 19. Peanuts as food for hogs were coming into favor by 1850. In sandy areas, hogs could be grazed on cultivated fields without injury to the soil. The editor of the *Albany Patriot* in 1846 recommended unharvested peanuts and potatoes as a substitute for the hand-feeding of corn. By this method he believed pork could be produced cheaper than Tennessee pork which was then selling for three and one-half cents a pound. *Soil of the South*, II (Sept. 1852), 333.
77. *Southern Cultivator*, I (Aug. 1843), 134; II (Dec. 1844), 206; IV (Jan. 1846), 17; XI (June 1853), 199.
78. N. P. Black, *Richard Peters*, 40.
79. *American Cotton Planter*, I (Nov. 1857), 350; *Southern Cultivator*, XVIII (April 1860), 130; XIX (Jan. 1860), 38.
80. *Ibid.*, XIX (July 1861), 218.
81. *Soil of the South*, I (Dec. 1851), 184; D. W. Lewis, *Transactions*, 46-48; *Southern Cultivator*, IX (Jan. 1851), 10.
82. *Ibid.*, X (Aug. 1852), 253; XI (March 1853), 126; XII (Aug. 1854), 287; D. W. Lewis, *Transactions*, 46.
83. N. P. Black Scrapbook, *loc. cit.*

CHAPTER X

1. D. W. Lewis, *Transactions*, 114, 118.
2. Emily P. Burke, *Reminiscences of Georgia*, 120-30; *Farmer's Register*, III (Aug. 1853), 246-47.
3. Born at Haverhill, New Hampshire, in 1781, Ward was educated at Middlebury College in Vermont and at Bowdoin where he graduated in medicine in 1821. He practiced this profession in Pennsylvania and Indiana before coming to Georgia. He resigned his chair at the University of Georgia in 1843 but continued to live in Athens, where he died on May 7, 1863. H. A. Kellar, *Solon Robinson*, II, 476; *Southern Cultivator*, XXI (July-Aug. 1863), 97.
4. H. A. Kellar, *Solon Robinson*, II, 475; *Atlanta Journal*, Dec. 27, 1936.
5. H. A. Kellar, *Solon Robinson*, II, 477; A. L. Demaree, *The American Agricultural Press*, 372.
6. *Southern Cultivator*, XI (May 1853), 153.
7. *Albany Cultivator*, I (Feb. 1844), 69.
8. *Augusta Chronicle*, July 12, 1800; *Southern Cultivator*, XIV (March 1856), 76; A. De Cardenc, *Grape Culture and Wine Making in the South* (Augusta, 1858), 7.
9. *Southern Cultivator*, III (Dec. 1845), 185.
10. *Southern Cultivator*, VI (June 1848), 88; VIII (June 1850), 88-89; XIV (June 1856), 76.
11. *Ibid.*, VIII (May 1847), 78; XVII (June 1850), 88-89.
12. *American Cotton Planter*, II (Dec. 1854), 381.
13. *Southern Cultivator*, XI (May 1853), 153; XII (Sept. 1854), 269.
14. *Ibid.*, III (Dec. 1845), 168; XII (Aug. 1854), 289.
15. Sir Charles Lyell, *A Second Visit*, 21.
16. *The Horticulturist*, II (Oct. 1847), 196.
17. *Soil of the South*, III (Feb. 1853), 451; IV (Feb. 1854), 58; *American Cotton Planter*, III (Nov. 1859), 359.
18. *Ibid.*, IV (Feb. 1854), 58.
19. Records of Bibb County Record Book "L," Clerk of Superior Court, 710; *Soil of the South*, IV (March 1854), 99.
20. *American Cotton Planter*, II (July 1854), 216.
21. *Southern Cultivator*, XIV (Dec. 1856), 373; XX (Jan. 1862), 1; *American Cotton Planter*, III (Jan. 1859), 24.

22. *Southern Cultivator*, XIV (March 1856), 91.
23. *Ibid.*, XIII (June 1855), 194.
24. *Ibid.*, XIV (March 1856), 91.
25. *Ibid.*, XII (Aug. 1854), 237.
26. Raphael Jacob Moses, Autobiography, an unpublished manuscript in the University of North Carolina Library, 64; Journal of Albert Moses Lauria, privately owned; *Georgia Citizen*, July 24, 1857.
27. *Southern Cultivator*, XVIII (July 1860), 219; (Oct. 1860), 309; XIX (Jan. 1861), 35.
28. *New York Daily Tribune*, Sept. 18, 1858, quoted in *Southern Cultivator*, XVI (Nov. 1858), 349.
29. Quoted in *American Cotton Planter and Soil of the South*, II (Nov. 1858), 347. Moses was among the few in Georgia who succeeded in growing a peach crop in 1847, a year of late frosts and freezing temperatures. He saved his crop by building fires in his orchard. *Georgia Citizen*, July 24, 1857.
30. R. J. Moses, Autobiography, *loc. cit.*, 64-65.
31. Mollie Van Buren Wood to Ruth Van Buren Tufts, Sept. 1, 1925, in private possession; Jarvis Van Buren, "Horticulture in the Interior of Georgia," *The Horticulturist*, VI (April 1851), 195; *Chronicle and Constitutionalist* (Augusta), March 11, 1885.
32. Records of Habersham County, Deed Book "HH," Clerk of Superior Court, 442; Slave Inhabitants of Habersham County, Schedule III, Census of 1860.
33. *The Horticulturist*, II (April 1851), 195; *Southern Cultivator*, XVIII (Aug. 1860), 245; Ruth Van Buren Tufts to James C. Bonner, Aug. 23, 1941.
34. *Southern Cultivator*, XVIII (Aug. 1860), 245.
35. *Soil of the South*, III (Dec. 1853), 761.
36. Andrew Jackson Downing, *The Fruits and Fruit Trees of America* (New York, 1859), 1 *et passim*. Cited hereafter as *Fruits and Fruit Trees*.
37. Patrick Barry, *The Fruit Garden* (New York, 1851), 1 *et passim*; *Southern Cultivator*, XV (March 1857), 94.
38. *Southern Cultivator*, VII (Feb. 1849), 28.
39. *Soil of the South*, II (Dec. 1852), 381.
40. *Southern Cultivator*, X (Sept. 1852), 271.
41. *Ibid.*, XI (Feb. 1853), 47; XVII (July 1859), 228; *Soil of the South*, III (Dec. 1852), 271.
42. William N. White, *et al.*, *Gardening for the South* (New York, 1868), 356; *Southern Cultivator*, XVII (July 1859), 280-281; A. J. Downing, *Fruits and Fruit Trees*, 124.
43. Kate Mason Rowland, ed., *The Poems of Frank O. Ticknor, M.D.* (Philadelphia, 1879), 121-22.
44. *Southern Cultivator*, XVII (July 1859), 220; (Sept. 1859), 280-81.
45. A. J. Downing, *Fruits and Fruit Trees*, 124; *Soil of the South*, III (Nov. 1853), 393.
46. *Southern Cultivator*, XII (Feb. 1854), 257, 269; XV (March 1857), 94.
47. *Soil of the South*, IV (Feb. 1854), 58.
48. *Southern Cultivator*, XI (Feb. 1853), 52; XVIII (Jan. 1860), 12; Armistead C. Gordon, Jr., "Francis Orray Ticknor," *Dictionary of American Biography*, XVIII, 525.
49. *South Countryman*, I (Feb. 1859), 58; *Southern Cultivator*, XVII (Feb. 1859), 79.
50. *Ibid.*, XII (June 1854), 189; XVII (Dec. 1859), 18; *Southern Field and Fireside*, II (Nov. 1860), 199.
51. *Southern Cultivator*, XVII (Dec. 1859), 18 (supplement); N. P. Black, *Richard Peters*, 46.
52. *Southern Field and Fireside*, II (May 1860), 7; (Nov. 1860), 197.
53. *Southern Cultivator*, IX (April 1851), 60; XV (April 1857), 120; XIX (May 1861), 170; Louis E. Berckmans, *Pear Culture in the South*, 11 (a brochure in the De Renne Collection, University of Georgia Library).

54. *Soil of the South,* IV (Feb. 1854), 58; *Southern Cultivator,* III (Nov. 1845), 168; Patrick Tailfer, et al., *A True and Historical Narrative of the Colony of Georgia,* 101-02.
55. *American Farmer,* X (Jan. 1829), 356; *Soil of the South,* III (Feb. 1853), 451; IV (Feb. 1854), 58; *American Cotton Planter,* III (Nov. 1859), 359.
56. *Colonial Records,* IV, 541; V, 500; Stephens' *Journal* (1741-1743), 95.
57. *Southern Cultivator,* III (Dec. 1845), 185; XIII (Oct. 1855), 320; *Colonial Records,* IV, 330.
58. *American Farmer,* VI (Feb. 11, 1825), 369.
59. *American Farmer,* VIII (March 1826), 83; *American Cotton Planter,* III (May 1859), 157-59.
60. *Southern Cultivator,* IV (Feb. 1846), 25; X (Feb. 1852), 57; XV (Sept. 1857), 281. The fine wines of these early vintners generally were not produced for market, but kept for home use and for special occasions. Returning home from the Civil War in 1865, B. L. Ridley spent the night at the home of John Bonner in Hancock County where his host gave him access to a store room of fine wines and brandies reserved for guests. "A sip from that forty-year old barrel was sweeter to me than the fruits of the Hesperides . . . or the nectar of the Gods," wrote Ridley. *Confederate Veteran,* III (Oct. 1895), 309.
61. IV (June 1846), 90.
62. *Proceedings of the Southern Vine Growers' Convention . . .,* 8 (a pamphlet published by the *Chronicle and Sentinel* at Augusta in 1860); *American Cotton Planter,* III (May 1859), 115.
63. A. De Cardenc, *Grape Culture,* 7; *Farmer's Register,* IV (Dec. 1836), 429; *Southern Cultivator,* X (March 1852), 75; The scuppernong was unlike the grapes from Europe and Asia in that the latter bore perfect flowers with stamens and pistils enclosed in the same corolla and capable of self-reproduction. *Ibid.,* XVII (Nov. 1859), 313; W. N. White, "Report upon Grapes," *The Horticulturist,* VII (Oct. 1857), 457.
64. A. De Cardenc, *Grape Culture,* 7,
65. *Ibid.;* W. N. White, *Gardening for the South,* 20.
66. *Southern Cultivator,* XVII (Dec. 1859), 361.
67. *The Horticulturist,* VII (Oct. 1857), 458; *Southern Cultivator,* XVI (March 1858), 94.
68. *The Horticulturist,* VII (Oct. 1857), 459-60; A. De Cardenc, *Grape Culture,* 6.
69. *American Farmer,* VI (Feb. 11, 1825), 369-70; *Southern Cultivator,* XIV (Feb. 1856), 63; XVI (March 1858), 94.
70. *Southern Cultivator,* XVIII (July 1860), 223.
71. *Ibid.,* XVI (March 1856), 94.
72. *Proceedings of the Southern Vine Growers' Convention,* 1, 11, 21.
73. *Ibid.,* 6, 8, 16.
74. *Southern Cultivator,* XIV (Oct. 1846), 315; XI (Sept. 1853), 280.
75. *Ibid.,* XV (Oct. 1855), 314; XVI Sept. 1856), 228.
76. *Ibid.,* XIV (Sept. 1856), 288; XIII (Oct. 1855), 314, 315, 320, 321; (Dec. 1855), 372; XIV (Dec. 1856), 377.
77. *Ibid.,* XIV (Oct. 1856), 315, 325; (Nov. 1856), 344.
78. *Ibid.,* XIV (Oct. 1856), 317, 376.
79. *Ibid.,* XV (Sept. 1857), 281; XVI (July 1858), 382.
80. *Ibid.,* XVII (Feb. 1859), 79; (April 1859), 116.
81. *Ibid.,* XVII (Oct. 1859), 302, 303; XVIII (Sept. 1860), 285. About this time Dimos Ponce reported a successful experiment in arresting rot in his vineyard in Hancock County by using a mixture of lime water and sulphur.
82. R. J. Moses, Autobiography, 2; Dennis Redmond, "Pomological Resources of the South," *Report of the Commissioner of Patents* (1858), 337-85; *New York Times,* Sept. 15, 1858; *New York Daily Tribune,* Sept. 18, 1858; *American Cotton Planter . . .,* III (June 1859), 19.

83. *South Countryman*, I (Jan. 1858), 18; *Southern Field and Fireside*, II (Aug. 1860), 95, 215; *Southern Cultivator*, XVII (Jan. 1859), 20.

84. *Ibid.*, XVIII (Jan. 1860), 1.

85. *Southern Field and Fireside*, II (Nov. 1860), 215.

CHAPTER XI

1. Frances S. Holmes, *The Southern Farmer and Market Gardener* (Charleston, 1842), 13; *Southern Cultivator*, X (July 1852), 207.
2. Frederick Law Olmsted, *Journey in the Seaboard Slave States* (New York, 1863), 583.
3. *Hancock Advertiser*, Feb. 25, 1828; *Soil of the South*, V (Jan. 1855), 24; *Southern Cultivator*, VI (June 1848), 87.
4. *Farmer's Register*, IV (April 1836), 763.
5. *Southern Cultivator*, XII (Sept. 1854), 258; *Soil of the South*, II (Dec. 1852), 381; III (Sept. 1853), 663.
6. *Ibid.*, I (Dec. 1851), 186; II (Dec. 1852), 381-82; III (Aug. 1853), 633; *American Cotton Planter . . .*, II (April 1858), 187; *Farmer's Register*, III (Aug. 1835), 246-47.
7. *Southern Cultivator*, XIV (March 1856), 76; XX (Sept.-Oct. 1862), 170.
8. *Ibid.*, XVIII (Jan. 1860), 37; *American Cotton Planter*, II (Nov. 1854), 349.
9. Mary Shepperson Crabb, George Alfred Peabody, an unpublished manuscript privately owned; *American Cotton Planter*, I (April 1857), 276-77.
10. *Ibid.*, I (Oct. 1857), 305; II (Dec. 1858), 377.
11. *American Cotton Planter*, II (April 1854), 120-21; *Southern Cultivator*, VII (Nov. 1849), 174.
12. A. J. Downing, *Fruits and Fruit Trees*, 684; *American Cotton Planter*, I (Sept. 1857), 276-77.
13. Robert Buchanan, *The Culture of the Grape and . . . Strawberry* (Cincinnati, 1853), 121; *Yearbook, U. S. Dept. of Agriculture* (1937), 448-50.
14. Robert Buchanan, *op. cit.*, 121, 128; *American Cotton Planter*, II (April 1854), 121-22.
15. *The Horticulturist*, VII (July 1857), 337.
16. M. S. Crabb, Charles Alfred Peabody, *loc. cit.*, 8; *Southern Cultivator*, V (Jan. 1857), 24; XIX (Feb. 1860), 53; A. J. Downing, *Fruits and Fruit Trees*, 684.
17. *Southern Cultivator*, X (Aug. 1852), 252; XII (Sept. 1854), 219.
18. *Soil of the South*, II (Dec. 1852), 381.
19. *Southern Cultivator*, XIV (Feb. 1856), 54; (June 1856), 188.
20. *Soil of the South*, I (Sept. 1851), 12; V (Feb. 1855), 56.
21. *Southern Cultivator*, XIV (June 1856), 188. The magnolia trees which flanked the avenue leading to the house of Dennis Redmond, later the property of the Augusta National Golf Club, were set out in the late 1850's by Prosper J. Berckmans who did much to stimulate the use of native evergreens for driveways. *American Cotton Planter*, I (Sept. 1857), 283.
22. *Soil of the South*, II (Dec. 1852), 381; III (Dec. 1853), 762.
23. *The Horticulturist*, IV (March 1850), 437.
24. W. N. White, "Live Fences," *Report of the Commissioner of Patents* (1855), 315; *DeBow's Review*, V (Jan. 1848), 83.
25. John C. Butler, *Historical Record of Macon and Central Georgia*, 339-40.
26. Quoted in George White, *Historical Collections of Georgia* (New York, 1854), 270, 271.
27. Mrs. David B. Canby to James C. Bonner, March 19, 1942; *Southern Cultivator*, XV (April 1857), 120; XVIII (June 1860), 190.
28. *Ibid.*, XIX (Nov. 1861), supplement; Loraine Meeks Cooney and Hattie

NOTES 229

C. Rainwater, eds., *Garden History of Georgia 1733-1933* (Atlanta, 1933), 62.
29. *Ibid.*, 132, 138.
30. *American Cotton Planter*, I (July 1857), 212; A. J. Downing, *A Treatise on the Theory and Practice of Landscape Gardening* . . . (New York, 1855), 57 (cited hereafter as *Landscape Gardening*).
31. *American Cotton Planter*, I (July 1857), 211, 212.
32. *Soil of the South*, II (July 1852), 212.
33. *American Cotton Planter*, I (July 1857), 211; *Southern Cultivator*, XV (March 1857), 86.
34. *Southern Field and Fireside*, II (Dec. 1860), 230; *American Cotton Planter*, I (Aug. 1857), 242.
35. George W. Featherstonehaugh, *Excursion Through the Slave States*, 2 vols. (London, 1844), II, 328-29.
36. *Southern Cultivator*, VI (Oct. 1848), 141; *South Countryman*, I (Feb. 1859), 61; H. A. Kellar, *Solon Robinson*, II, 475-76.
37. *Southern Cultivator*, VI (Oct. 1848), 141-42. At an early date pecans were grown as a crop in the lower Mississippi Valley. A Texan, writing from Kerr County in 1857 stated that 200,000 bushels of the nuts were exported annually from that state and expressed surprise that they were not grown in other Southern states. The abundance of other native nut-bearing trees, such as walnuts and chestnuts, may explain the late entry of pecans to the list of money crops in the old cotton belt. L. C. Gray, *op. cit.*, II, 460; *The Horticulturist*, VII (July 1857), 330.
38. *Soil of the South*, II (Jan. 1853), 12.
39. Emily Burke, *Reminiscences*, 320.
40. *Southern Cultivator*, III (May 1845), 75.
41. *Soil of the South*, I (Dec. 1851, extra edition), 75.
42. *Southern Field and Fireside*, I (Dec. 1859), 247.
43. *American Cotton Planter*, I (Jan. 1857), 49.
44. *Prairie Farmer*, XIII (Jan. 1853), 2-3; A. L. Demaree, *The American Agricultural Press*, 71, 263, 265; A. J. Downing, *Cottage Residences*, 32-33, 333; A. P. Downing, *Landscape Gardening*, 374-75; Howard Major, *The Domestic Architecture of the Early American Republic* (Philadelphia, 1926), 24 et passim; *American Agriculturist*, XVII (March 1858), 73; *Southern Cultivator*, VI (Oct. 1848), 141.
45. *Ibid.*, IV (Dec. 1846), 362; VI (June 1848), 89; (Oct. 1848), 154; XVII (Aug. 1860), 245.
46. *Soil of the South*, I (Aug. 1857), 124-25.
47. *Ibid.*, II (Jan. 1853), 104; *Southern Cultivator*, XIV (Dec. 1846), 362.
48. George R. Fairbanks, "Farmer's Homes," *Soil of the South*, II, (Feb. 1852), 29.
49. C. Reagles, "The Philosophy of Suburban Cottage Homes," *Southern Cultivator*, XV (Nov. 1857), 340.
50. *Soil of the South*, I (Jan. 1851), 10; II (April 1852), 253.
51. *Southern Cultivator*, XII (Jan. 1854), 30.
52. *Ibid.*, XVI (Oct. 1858), 307; *South Countryman*, I (Feb. 1859), 61; H. A. Kellar, *Solon Robinson*, II, 482; Fredrika Bremer, *The Homes of the New World*, 2 vols. (New York, 1853), I, 330-46, 370.
53. M. S. Crabb, Charles Alfred Peabody, *loc. cit.*, 1 et passim; Emily G. Ravenel to James C. Bonner, May 3, 1942, in possession of the author.
54. *Southern Cultivator*, XVIII (Oct. 1860), 301; Ralph B. Flanders, "Farish Carter, A Forgotten Man of the Old South," *Georgia Historical Quarterly*, XV (March 1931), 147; R. J. Moses, Autobiography, 141.
55. F. L. Olmsted, *Journey in the Seaboard Slave States*, 537.
56. *Soil of the South*, IV (March 1854), 103.
57. *Southern Cultivator*, XXII (Jan. 1864), 15.
58. *Soil of the South*, V (Aug. 1855), 270.
59. *Southern Cultivator*, X (Aug. 1852), 238, 252.
60. John M. Turner, "Negro Houses,"

Ibid., XV (June 1857), 170; *Soil of the South*, II (March 1852), 229.
61. *Ibid.*, I (Jan. 1851), 13.
62. *Ibid.*, I (Jan. 1851), 13. Dimos Ponce had a reputation for curing hams of unusual flavor and quality and they brought top prices in the Savannah market. His process involved the use of salt, saltpetre, and brown sugar, and the hams were never marketed until properly aged. The curing property of heat from smoke and of age were not generally recognized, since most Georgia hams were consumed within the first year after curing began. *Southern Cultivator*, XVII (Dec. 1859), 358; D. W. Lewis, *Transactions*, 322.
63. *Southern Cultivator*, XVIII (Aug. 1860), 262; *Soil of the South*, I (Jan. 1851), 14.
64. *Albany Cultivator*, IV (May 1847), 160; *Southern Cultivator*, XII (Sept. 1854), 293.
65. *Ibid.*, X (Dec. 1852), 365.

CHAPTER XII

1. *Southern Cultivator*, II (May 1844), 83.
2. *Ibid.*, XI (June 1853), 168.
3. *Ibid.*, XVII (June 1859), 163; J. R. Cotting, *Essay on Soils*, 24; *South Countryman*, I (Feb. 1859), 57; (April 1859), 82.
4. John R. Cotting, *Report of a Geological and Agricultural Survey of Burke and Richmond Counties, Georgia* (Augusta, 1836), 20; J. R. Cotting, *Essay on Soils*, 51; *Southern Cultivator*, II (May 1844), 75.
5. *Ibid.*, XI (Jan. 1853), 8; *Southern Field and Fireside*, II (Dec. 1860), 246.
6. Edmond M. Pendleton, "Compost Manure," D. W. Lewis, *Transactions*, 373-75; *Soil of the South*, I (March 1851), 52; *Southern Cultivator*, XV (March 1857), 78.
7. *Ibid.*, VIII (March 1850), 35.
8. *Ibid.*, XI (Dec. 1853), 355; XIII (March 1855), 76.
9. *Soil of the South*, I (April 1851), 59.
10. *Southern Cultivator*, XI (Feb. 1853), 39; XVII (March 1859), 75.
11. *Soil of the South*, III (Feb. 1853), 425; *Southern Cultivator*, XI (March 1853), 82.
12. *Ibid.*, XI (March 1853), 73.
13. *Ibid.*, XI (Feb. 1853), 45; (April 1853), 105; XVII (Feb. 1859), 75.
14. *Ibid.*, XI (March 1853), 105; XVII (July 1859), 202; *Southern Field and Fireside*, II (Oct. 1860), 174.
15. *The Advocate* (Carrollton, Ga.), Oct. 26, 1860; *Southern Cultivator*, XVIII (July 1859), 202.
16. *American Cotton Planter*, II (Nov. 1858), 344.
17. *Southern Field and Fireside*, II (Oct. 1860), 182; *Southern Cultivator*, XIX (Feb. 1861), 43.
18. *Ibid.*, XVII (Dec. 1859), 345; XVIII (Nov. 1860), 301; *Southern Field and Fireside*, II (Oct. 1860), 182; R. B. Flanders, "Planters' Problems in Ante-Bellum Georgia," loc. cit., 35; George F. Hunnicutt, ed., *David Dickson's and James M. Smith's Farming* (Atlanta, 1910), 201 (cited hereafter as *David Dickson*).
19. *Ibid.*, 18; *Southern Cultivator*, XVIII (Jan. 1860), 16-17; (Oct. 1860), 203; *South Countryman*, I (Jan. 1859), 27; Productions of Agriculture in Hancock County, Schedule IV, Census of 1850.
20. G. F. Hunnicutt, *David Dickson*, 15; *Southern Cultivator*, XVIII (Jan. 1860), 16-17.
21. *Ibid.*, XVII (Aug. 1859), 255; XVIII (June 1860) 174, 175; (July 1860), 211; (Nov. 1860), 301; *Southern Field and Fireside*, II (Oct. 1860), 182; David Dickson, *Improved Cotton*, a broadside published at Oxford, Georgia, Oct. 1, 1858; *South Countryman*, I (Feb. 1859), 68. For the best discussion available on cotton breeding in the ante-bellum South, see John H. Moore, *Agriculture in Ante-Bellum Mississippi* (New York, 1958), 145-60.
22. *Ibid.*, I (Jan. 1859), 27; Productions of Agriculture in Hancock County, Schedule IV, Census of 1860; *South-*

ern Cultivator, XVII (Dec. 1859), 345, 368; XVIII (July 1860), 203; R. B. Flanders, "Planters' Problems in Ante-Bellum Georgia," loc. cit., 35.
23. Southern Cultivator, XVII (April 1859), 125; (Sept. 1859), 261; XVIII (Jan. 1860), 17-18.
24. Ibid., XVII (Sept. 1859), 261-262; XVIII (Nov. 1860), 339-340; XIX (April 1861), 113; Soil of the South, V (Nov. 1855), 321.
25. David A. Wells, Yearbook of Agriculture (1856), 22; Southern Cultivator, XVII (July 1859), 200; American Cotton Planter, II (March 1859), 97.
26. Southern Cultivator, XIV (Feb. 1856), 52; (Dec. 1856), 333; XVIII (Nov. 1860), 339.
27. Ibid., XVI (Nov. 1858), 337.
28. Savannah Chamber of Commerce to Tuttle H. Audas, July 3, 1843, in Records of the Hancock Planters' Club, Office of County Clerk, Sparta, Georgia; Southern Recorder (supplement, June 4, 1850), 337; Southern Cultivator, V (Sept. 1847), 136; XIV (Dec. 1856), 337.
29. Soil of the South, II (March 1852), 69.
30. Southern Cultivator, III (Feb. 1845), 25; XI (June 1853), 161.
31. Dimos Ponce to Tuttle H. Audas, Oct. 31, 1844, Records of the Hancock Planters' Club, loc. cit.; Minutes, 1844, ibid.
32. Southern Cultivator, III (Feb. 1845), 26.
33. Ibid., III (Feb. 1845), 42, 43.
34. Rome Courier, July 24, 1851.
35. Southern Cultivator, XX (April 1862), 83; Acts of the General Assembly (1862), P. 5.
36. D. W. Lewis, Transactions, 94; Farmer's Register, IV (Dec. 1835), 454; IV (April 1836), 763.
37. Southern Cultivator, XVIII (Oct. 1859), 303.
38. South Countryman, I (Jan. 1859), 27; Southern Cultivator, VIII (July 1850), 99; XIX (Jan. 1851), 19; G. F. Hunnicutt, David Dickson, 96-101; H. A. Kellar, Solon Robinson, II, 478; James M. Chambers, "An Essay on the Treatment and Cultivation of Corn," D. W. Lewis, Transactions, 102.
39. Ibid., 203; E. M. Pendleton, Textbook of Agriculture, 137; American Cotton Planter, I (Jan. 1857), 17; (June 1857), 158, 164; Southern Cultivator, XIX (April 1861), 113; XXII (March 1864), 51, 52.
40. D. W. Lewis, Transactions, 208; Southern Cultivator, VIII (July 1850), 99; XI (Jan. 1853), 9; XII (Feb. 1854), 43; XII (Aug. 1854), 243; XVIII (March 1860), 98; XXII (March 1864), 42, 51; American Cotton Planter, I (June 1857), 154, 164.
41. U. B. Phillips, Life and Labor, 177; South Countryman, I (May 1859), 160.
42. Federal Union, Jan. 17, 1860; Southern Cultivator, XVII (Sept. 1859), 276.
43. Southern Agriculturist, I (Dec. 1828), 523.
44. Charles Colcock Jones, Jr., Negro Myths From the Georgia Coast, Told in the Vernacular (Boston, 1888), 138-40.
45. Southern Cultivator, XVIII (March 1860), 98; American Cotton Planter, III (July 1859), 356.
46. Ibid., III (Feb. 1859), 67, 68; (July 1859), 228-29.
47. Ibid., II (Nov. 1858), 355; III (Feb. 1859), 67-68; (June 1859), 197; Southern Cultivator, VIII (Nov. 1850), 162.
48. Ibid., VIII (Nov. 1850), 162; XIX (March 1861), 73; H. S. Randall, Sheep Husbandry, 155.
49. Southern Cultivator, VIII (Nov. 1850), 162.
50. Soil of the South, III (Aug. 1853), 650.
51. Southern Cultivator, XI (Aug. 1853), 227.
52. Ibid., VIII (Nov. 1850), 162.
53. Ibid., XXII (March 1864), 48.
54. Ibid., XVIII (Sept. 1860), 276-77.
55. Ibid., VIII (Nov. 1850), 162; XI (Oct. 1853), 302; XV (March 1857), 170; Farmer's Register, III (Dec. 1835), 494-95.
56. Joseph R. Wilson, Mutual Relations

of Masters and Slaves, 20; Soil of the South, III (Feb. 1853), 458-59; Southern Cultivator, XIX (April 1861), 105-10.
57. Minute Book "C," Records of Early County, 286-89.
58. Southern Cultivator, VIII (Nov. 1850), 162; XVIII (Sept. 1860), 276-77; (Oct. 1860), 304, 305.
59. Ibid., XI (Oct. 1853), 301.
60. Ibid., VII (March 1849), 68; XVIII (Jan. 1860), 17-18.
61. Ibid., XV (June 1857), 170; (Aug. 1857), 237.
62. Ibid., XVII (Nov. 1859), 345, 368; XVIII (Feb. 1860), 44; (Nov. 1860), 301; XIX (April 1861), 113; G. F. Hunnicutt, David Dickson, 30, 32, 35; R. B. Flanders, Plantation Slavery in Georgia, 35.
63. Southern Cultivator, II (June 1844), 97; VI (Sept. 1848), 134; VII (May 1849), 75; VIII (Sept. 1850), 135; XIV (July 1856), 209; XVIII (Feb. 1860), 44; American Cotton Planter, IV (May 1854), 149.
64. Southern Cultivator, III (Feb. 1845), 26; VIII (April 1850), 50; XVIII (Sept. 1860), 287.
65. Ibid., XX (July-Aug. 1862), 287.
66. Ibid., IV (July 1846), 106; XV (March 1855), 76; XIV (Nov. 1856), 339.
67. Southern Field and Fireside, II (Dec. 1860), 222-27; H. A. Kellar, Solon Robinson, II, 479. Robinson was born in Connecticut in 1803 and later moved to Indiana where he became an outstanding commentator on American rural life. He was connected at various times in the editorial work with the American Agriculturist, The Plow, and the New York Tribune. Ibid., I, 3-4.

Selected Bibliography

NOTES ON SOURCES

THE ONLY book which has attempted to give the subject of Georgia agricultural history any degree of comprehensive treatment is Willard Range, *A Century of Georgia Agriculture, 1850-1950* (Athens, 1954). For the period before 1860 Lewis Cecil Gray, *History of Agriculture in the Southern United States to 1860* (2 vols., New York, 1941) contains much general information on Georgia agriculture. This subject was introduced as a field of research in 1929 when Professor E. Merton Coulter published in *Agricultural History* (III, Oct., 1929) a short paper entitled "A Century on a Georgia Plantation." Since that time a dozen articles on early Georgia agriculture have appeared in historical journals.

Perhaps the most valuable single source of information on the earliest period of Georgia agriculture is *The Colonial Records of the State of Georgia* (16 vols., Atlanta, 1904-1916), in which appear numerous accounts having particular reference to agricultural conditions. The most valuable of these is William Stephens, "A Journal of the Proceedings in Georgia" (Vol. IV) which covers the period from October 20, 1737 to October 4, 1749. The remainder of Stephens' journal covering the period through 1745 has been edited by E. Merton Coulter and published in two volumes by the University of Georgia Press in 1958 and 1959. In *Collections of the Georgia Historical Society* (11 vols., Savannah, 1840-1955) and in Allen D. Candler, ed., *The Revolutionary Records of the State of Georgia* (3 vols., Atlanta, 1908) are scattered but highly useful references to early agriculture in the Georgia colony. Among the official manuscript archives of the Georgia Department of Archives and History in Atlanta are a number of useful materials such as Colonial farm inventories, records of land distribution, register of early cattle brands, and records on administrations of Colonial estates. The libraries of the University of Georgia, the Georgia Historical Society at Savannah, the University of North Carolina, and Duke University are particularly noted for their unofficial manuscripts on Southern history, many of which deal with various phases of Georgia agriculture before 1860. The Phillips and the De Renne collections of the University of Georgia and the Joseph Clay Collection of the Georgia Historical Society are particularly noteworthy. Important early material appearing in print and having unusual value in a study of Colonial agriculture include Eliza Lucas, *Journal and Letters of Eliza Lucas* (Wormsloe, 1850); William Bartram, *Travels Through North and South Carolina and Georgia* . . . (Philadelphia, 1791); Thomas Stephens, *The Hard Case of the Distressed People of Georgia* (London, 1742); Thomas Boreman, *A Compendious Account of the Whole Act of Breeding, Nursing and Right Ordering of the Silkworm* (London, 1733); and John G. W. DeBrahm, *History of the Province of Georgia* (Wormsloe, 1849).

Materials on agricultural development covering the period from the end of the Revolution to the opening of the nineteenth century are scarcer than for any other twenty-year period in the state's history. The problems of land organization and distribution in this period are adequately treated in Milton Sydney Heath, *Constitutionalism Liberalism: The Role of the State in the Economic Development of Georgia to 1860* (Cambridge, 1954) and S. Guyton McLendon, *History of the Public Domain of Georgia* (Atlanta, 1924). The more important travel accounts for this period of agricultural history include J.F.D. Smyth, *A Tour of the United States of America* (2 vols., London, 1784) and William Lee, *The True and Interesting Travels of William Lee* . . . (New York, 1782). Records in the office of the Surveyor General of Georgia and the land records of the Department of Archives and History, to which the former are attached, are complete and well organized. They provide a wealth of valuable minutiae on this subject. The official archives of several of the older counties of Georgia, notably Hancock, Greene, Richmond, Jefferson, and Liberty, contain many farm and plantation inventories for the period which are essential for a knowledge of agricultural productions, livestock, and equipment. Important printed sources possessing valuable information are William A. Stevens, *A History of Georgia* (2 vols., New York, 1847); Denison Olmsted, *Memoir of Eli Whitney* (New Haven, 1846); Robert and George Watkins, eds., *Digest of the Laws of the State of Georgia . . . to 1798* (Philadelphia, 1800); and Charles H. Haskins, "The Yazoo Land Companies," in *Papers of the American Historical Association* (Vol. 80, New York, 1891).

Significant references to agricultural life in Georgia from 1800 to 1840 are found in various travel accounts, the most important of which are John Melish, *Travels Through the United States of America in the Years 1806, 1807, and 1811* (2 vols., Philadelphia, 1815); James Stuart, *Three Years in North America* (2 vols., New York, 1833);

Tyrone Power, *Impressions of America During the Year 1833-1834, and 1835* (2 vols., London, 1836); and Mrs. Anne Royall, *Mrs. Royall's Southern Tour, or the Second Series of the Black Book* (2 vols., Washington, 1831). Of great significance are the early agricultural journals, two of which were published outside the South yet containing many references to Georgia agriculture. The most important of these is the *Southern Agriculturist* (19 vols., Charleston, 1828-1846), for which Thomas Spalding of Sapelo Island contributed a large number of articles on Georgia. Others are the *Farmers' Register* (11 vols., Shellbanks and Petersburg, 1833-1843); the *American Farmer* (21 vols., Baltimore and Washington, 1819-1839); the *Genesee Farmer* (9 vols., Rochester, New York, 1831-1839); and *The Cultivator* (32 vols., Albany, New York, 1834-1865). Georgia newspaper files began in 1763 but these did not become a significant source of agricultural history until the second or third decade of the nineteenth century. Important among these later newspapers are the *Hancock Advertiser* (1825-1840); the *Georgia Journal* which began publication at Milledgeville in 1808; the *Southern Recorder* which began in 1820; and the *Federal Union* which entered the field in 1830. All were weeklies and they all represented the rapidly developing Cotton Belt. *Niles' Weekly Register* (76 vols., Baltimore and Philadelphia, 1811-1849) is also a significant source for this period. Important printed primary sources are Ulrich B. Phillips et al., eds., *A Documentary History of American Industrial Society* (10 vols., Cleveland, 1910). Valuable manuscript materials in the form of letters, plantation records and diaries are found in the libraries of Duke University, the University of North Carolina, and the University of Georgia. Complete and carefully catalogued records on the various land lotteries from 1803 to 1832 are in the Georgia Department of Archives and History. There are no comprehensive materials on Creek and Cherokee Indian agriculture for this period. However, scattered references to these subjects are found in such works as Rufus Anderson, *Memoir of Catherine Brown* (New York, 1827); Jedidiah Morse, *A Report to the Secretary of War on Indian Affairs* (New Haven, 1822); and in Henry R. Schoolcraft, ed., *Information Respecting...the Indian Tribes of the United States* (6 vols., Philadelphia, 1855).

For the twenty-year period following 1840 there is a wealth of source material on Southern agricultural history. Manuscripts and official documents and the printed works of both a primary and secondary nature are especially numerous for the Georgia area. The most valuable single source of information is the *Southern Cultivator* (1843-1935). Published originally at Augusta it was the spokesman of the entire cotton-growing region. The Georgia agricultural press was augmented in 1851 by the *Soil of the South* at Columbus. It continued until 1856 when it merged with the *American Cotton Planter* at Montgomery. The *South Countryman* (Marietta, 1859-1860) and the

Southern Field and Fireside (Augusta, 1859-1864) complete the list. The last was an eclectic journal including various interests, but it possessed a definite agricultural bias. Beginning in 1850 three schedules were made up by census enumerators which have become invaluable documents for a study of agricultural history. These are Schedules I, II, and IV; that is, "Free Inhabitants," "Slave Inhabitants," and "Productions of Agriculture," respectively. Of these, the last is the most useful document. When collated with the other two, an average of approximately fifty items of information can be assembled on each individual farmer whose name appeared on the schedules. These original records for Georgia are on microfilm at the National Archives. The most important printed source for this period of Georgia agriculture are David W. Lewis, ed., *Transactions of the Southern Central Agricultural Society, 1846-1851* (Macon, 1852); Herbert A. Keller, ed., *Solon Robinson, Pioneer and Agriculturalist* (2 vols., Indianapolis, 1936); Henry S. Randall, *Sheep Husbandry in the South* (Philadelphia, 1848); Emily P. Burke, *Reminiscences of Georgia* (Oberlin, 1850); John Ruggles Cotting, *An Essay on the Soils and Available Manures of the State of Georgia* (Milledgeville, 1843); Joseph Jones, *First Report to the Cotton Planters' Convention of Georgia* ... (Augusta, 1860); Joseph Jones, *The Agricultural Resources of Georgia* (Augusta, 1861); Louis E. Berckmans, *Pear Culture in the South* (Augusta, 1859); Prosper J. A. Berckmans, *A Great Fruit Country* (Augusta, n.d.); William L. Black, *A New Industry, or Raising the Angora Goat, and Mohair, for Profit* (Fort Worth, 1900); John Lord Hayes, *The Angora Goat, its Origin, Culture, and Products* (Boston, 1868); George F. Hunnicutt, ed., *David Dickson's and James M. Smith's Farming* (Atlanta, 1910); Dennis Redmond, *Sorgho Sucre, or, Chinese Sugar Cane*... (Augusta, 1856); Charles Wallace Howard, *A Manual of the Cultivation of the Grasses and Forage Plants at the South* (Atlanta, 1881); and David Lewis Phares, *The Farmer's Book of Grasses and Other Forage Plants, for the Southern United States* (Starkville, Mississippi, 1881). The more important scholarly works are E. Merton Coulter, *Thomas Spalding of Sapelo* (Baton Rouge, 1940), which contains much information on agricultural conditions on the Georgia coast; Ralph Betts Flanders, *Plantation Slavery in Georgia* (Chapel Hill, 1933); and Ulrich Bonnell Phillips, *Life and Labor in the Old South* (Boston, 1929), both of which give excellent treatments of the institution of slavery in Georgia.

Most of this study came from widely scattered and unclassified sources, all of which appear in the footnotes.

Index

Adams, Benjamin F., 85
Affleck, Thomas, 129, 130, 203
Agricultural journals, 119-123; *Albany Cultivator*, 94; *American Agriculturist*, 84; *American Cotton Planter*, 97, 105, 121, 153, 170; *Farmers' Register*, 75, 97, 195; *Genesee Farmer*, 120, 129; *The Horticulturist*, 171; *Soil of the South*, 91, 120-121, 153, 170, 177, 182, 183; *South Countryman*, 119-122, 132; *Southern Agriculturist*, 90; *Southern Cultivator*, 79, 82, 83, 91, 97, 104, 106-109, 111, 119-122, 132, 145, 150, 152, 155, 159, 160, 164, 179, 180, 182, 199; *Southern Field and Fireside*, 120
Agricultural schools, 123-126
Agricultural Societies, 110-119; Agricultural Association of Georgia, 116; Agricultural Club of Lower Georgia, 115; American Pomological Society, 118; Cass County Society, 115; Central Horticultural Society (Macon), 113, 117; Cincinnati (Ohio) Horticultural Society, 171; Etowah Agricultural and Mechanical Association (Rome), 92, 115; Georgia Silk and Agricultural Society, 220n; Horticultural Society of Georgia, 118; Mechanics Institute of Georgia, 118; Muscogee-Russell Agricultural Society (Columbus), 115; New York Agricultural Society, 139; Pendleton (S.C.) Agricultural Society, 110; Planters Club of Hancock County, 112-115; Southern Central Agricultural Society, 112, 116-119, 123, 156; U. S. Agricultural Society, 85; Union Agricultural Society (Darien) 110;

Wilkes County Agricultural Society, 115; others, 219n
Alabama, 41, 42, 63, 69, 71, 72, 82, 83, 98, 120, 153, 164
Allen, Wyatt, 79
Anderson, Joseph R., 91
Anderson, Thomas, 54
Andrews, Garnett, 67
Anthony, Bolling, 54
Apples, 55, 155-158, 166
Architecture, 178-185
Atlanta Intelligencer, 165
Augusta Chronicle, 141
Augusta Constitutionalist, 154
Axt, Charles, 164-165

Bachman, Dr. John, 136
Baer, Charles, 122, 123
Bakers, Benjamin, 7
Baltimore Daily Times, 83
Bancroft, Edward, 154
Barker, Joseph, 26
Barnes, V. M., 146
Barry, Patrick, 156
Bass, Nathan, 118
Battey, George M., 147, 148
Battey, Dr. Robert, 83, 84, 169, 182
Baxter, Eli H., 63, 112
Beal, Zephemiah, 50
Bell, Hiram P., 44, 45
Berckmans, Louis E., 118, 159, 173
Berckmans, Prosper Jules, 156, 159, 162, 163, 173, 174
Berrien, John M., 113, 148
Berrien, Thomas M., 113
Board of Agriculture and Rural Economy, 111

Bolzius, Rev. John Martin, 4, 23, 24, 27
Bonner, John, 84, 145, 146, 147, 182, 191, 227n
Bonner, Zadock, 81
Bounty grants, 32-34
Bouverie Farm, 3
Brown, Algernon S., 112
Buckner, Col. Singleton, 158
Bulloch, Archibald, 11
Burke, Emily, 43, 68, 87, 177
Butler, Major, 49
Butler, Pierce, 196
Butler's Island, 196

California, 143, 144, 158
Callaway's Iron Works, 54
Camak, James, 113, 119, 120, 150, 160, 169, 194, 195
Carmichael, John C., 83
Carter, Farish, 87, 181
Cattle industry, 25-30, 135
Cedar Valley, 70, 101, 107
Chambers, James M., 120, 196
Chambers, William H., 177
Cheese-making, 136
Cherokee Indians, 8, 37, 39, 40, 43, 44, 45, 46, 155, 157
Cherokee Georgia, 44, 65, 66, 69, 112, 115
Chinese pea, 86, 102
Chinese poultry, 148
Chinese sorghums, 83, 84.
Chivers, Thomas H., 75
Clarke, Elijah, 32
Cities and Towns: Abbeville, S.C., 165; Aerschot, Belgium, 173; Aiken, S.C., 163; Atlanta, 40, 109, 118, 130, 134, 143, 148, 151, 158, 164, 165, 184; Athens, 73, 100, 118, 119, 154, 172, 176, 194; Augusta, 1, 8, 9, 22, 33, 36, 41, 46, 48, 49, 50, 54, 55, 69, 80, 82, 83, 86, 90, 111, 118, 119, 120, 122, 127, 129, 137, 153, 159, 162, 163, 164, 165, 170-76, 183, 184, 192; Baltimore, Md., 36, 123, 125; Boston, Mass., 83; Bremen, Germany, 66; Buchanan, 70; Carrollton, 42; Cartersville, 91, 107; Cassville, 140; Cedartown, 70, 101; Chambersburg, Pa., 182; Charleston, S.C., 22, 131, 136, 152, 155, 172; Chicago, Ill., 170; Cincinnati, Ohio, 158, 165, 171; Clarksville, 45, 46, 118, 136, 137, 155, 163; Columbia, S.C., 162; Columbus, 41, 42, 46, 64, 68, 69, 71, 80, 83, 91, 115, 120, 121, 123, 128, 154, 158, 163, 166, 169, 170, 174, 175, 181, 184, 197-99; Covington, 98, 158; Crawfordville, 164, 165; Cumming, 45; Dahlonega, 44, 45; Dalton, 165; Darien, 4, 21, 49, 110; Eatonton, 57, 91, 111, 130; Ebenezer, 1, 16, 22, 23, 25, 26, 28, 29; Edenborough, 50; Federaltown, 50; Franklin, N.C., 157; Frederica, 1, 21, 30; Fredericksburg, Va., 108; Germantown, Pa., 134; Greensboro, 90, 130, 188; Hamilton, 194; Harrisburg, 50; Kingston, 115, 120, 174, 181; Knoxville, Tenn., 104; Lexington, 58, 88, 110; Louisville, 36; Macon, 41, 43, 46, 63, 68, 69, 81, 98, 113, 116-119, 121, 123, 125, 148, 149, 152, 154, 166, 169, 173, 182, 184, 188; Madison, 45, 124; Madison's Parish, La., 106; Mallorysville, 166; Manassas, Va., 108; Marietta, 108, 123, 124; Marthasville, 134; Midway, 1; Milledgeville, 42, 46, 66, 69, 83, 57, 62, 64, 111, 116, 117, 122, 130, 158, 162, 179, 193; Milliken's Bend, La., 106; Mont Vale Springs, Tenn., 104; Montezuma, 154; Montgomery, Ala., 120, 152, 153, 158, 165, 170; Montpelier, 50; Mount Zion, 193; Natchez, Miss., 69; Nashville, Tenn., 66; New Orleans, 78, 83; New York, 56, 66, 82, 94, 118, 120, 121, 129, 139, 140, 145, 154-157, 166, 170; Newborn, 130; Newnan, 121, 174; Norfolk, Va., 155; Petersburg, 50; Philadelphia, Pa., 36, 84, 154, 171; Pittsburgh, Pa., 96; Platteville, Iowa, 144; Purysburg, S.C., 27, 28; Quitman, 144; Raytown, 94; Richmond, Va., 82, 108, 155; Rocky Mount, N.C., 54; Rome, 37, 44, 84, 115, 148, 160, 169; Ruckersville, 79; St. Louis, Mo., 143, 170; St. Paul's Parish, 162; Sand Town, 108, 109; Sandersville, 42, 114; Savannah, 1, 3, 4, 7-9, 13, 15, 16, 21, 23, 26, 30, 43, 48, 49, 52, 53, 55, 57, 82, 87, 110, 115, 137, 152, 154, 155, 158, 163, 168, 188, 191, 192; Sparta, 42, 60, 69, 112, 116, 127, 128, 162, 188, 194; Spring Place, 158; Springfield, Ill., 85; Standfordville, 62; Staunton, Va., 108; Stone Mountain, 116; Talbotton, 69; Thomaston, 125; Union Point, 163; Utica, Miss., 104; Vernonburg, 7; Warrenton, 162; Washington, 130, 165; Watkinsville, 100; Waynesboro, 87, 140; Weathersfield, Conn., 169; West Point, 72; White Plains, 158; Winfield, 146; Woodbury, Conn., 170

INDEX

Clayton, Judge Augustin S., 100
Cloud, Dr. Noah B., 101, 107, 120, 153
Cobb, Howell, 118, 119
Coleman, James L., 90
Collins, Charles, 147
Columbus Enquirer, 121
Cooper, Mark A., 91, 116, 118
Cooper, Thomas, 36
Coram, Thomas, 7
Corn, 21, 194-196
Cottage Residences (Downing), 178
Cotting, Dr. John Ruggles, 111
Cotton, sea island, 51-52; upland, 49-56, 57, 76-77, 186, 191-194
Couper, John, 150, 163
Cowboys (and "cow hunters"), 26, 222n; "pindars", 26-29
Crawford, George W., 91, 116
Crawford, Joel, 112, 162, 200
Crawford, William H., 158
Creek Indians, 8, 37, 40, 41, 43, 44, 72
Croom, Isaac, 101
Cumberland Island, 51
Cunningham, John, 90, 130, 131, 188

Davidson, Paul, 94
Davis, I. N., 121
Davis, James B., 142
d'Estaing, Count, 32
Devereaux, Samuel M., 162
Dickson, David, 85, 87, 96, 101, 112, 113, 125, 189, 190, 191, 195, 201
Dickson, Thomas J., 113, 195
Diversification, 73, 92
Douglas, David, 22
Downing, Andrew J., 156, 174, 178

Early, Joel, 51
Earnest, Asa E., 75
East Indies, 47, 148
Elliott, Stephen, 81, 113, 117, 118, 149
Emigration, 42, 61-72
European agriculture, 97, 186
European grapes, 159
Eve, Judge John P., 115
Eve, William J., 127

Farrar, John, 130
Farrow, John, 62, 97
Featherstonehaugh, George W., 42
Federal Union, 100, 196
Fertilizers, 187-191, 195
Fields, John M., 166
Fish, George W., 152
Fish ponds, 83

Fitzwalter, Joseph, 30
Florida, 8, 27, 40, 42, 63, 78, 144; Florida Panhandle, 63
Flowers and shrubs, 80, 172-175
Food crops, 21-23
Forsyth, Governor John, 58, 68, 179
Fort, Arthur, 53
Free labor, 4, 89, 91, 109
French burr-millstones, 82; rice, 48
Fruit and Fruit Trees of America (Downing), 156
Fruit Garden, The (Barry), 156

Gardens, 168-172
Georgia Counties: Appling, 39; Baldwin, 57, 59, 87, 181; Bartow, 138; Bibb, 75; Burke, 48, 49, 57, 113, 131, 139, 202; Butts, 95; Campbell, 72, 108; Carroll, 41, 65, 72, 81; Cass, 44, 65, 66, 88, 106, 107, 115, 131, 134, 141, 156, 181; Chatham, 111, 113; Chatooga, 65, 66; Cherokee, 44; Clarke, 68, 94, 95, 100, 113, 128, 157; Cobb, 44, 109; Coweta, 41, 72; Dooly, 109; Early, 39, 112, 200; Effingham, 36, 48; Elbert, 87; Fayette, 41; Floyd, 65, 101, 106, 113, 132, 180; Forsyth, 44; Franklin, 32, 33, 35, 36, 48; Gilmer, 81; Gordon, 65, 69, 134, 143; Greene, 48, 50, 51, 130, 145, 152, 182; Gwinnett, 39; Habersham, 39, 41, 45, 77, 113, 155, 156, 157; Hall, 39, 81, 157; Hancock, 42, 56, 63, 67, 76, 78, 81, 87, 88, 97, 98, 100, 112, 113, 114-116, 123, 130, 131, 146, 147, 150, 151, 162, 189-191, 193, 194, 201; Harris, 41, 65, 72, 193; Heard, 65, 72; Henry, 125; Houston, 98; Irwin, 39, 43; Jackson, 58, 157; Jasper, 57; Jefferson, 130, 138; Laurens, 57, 58, 159; Lee, 72, 85; Liberty, 36, 187; Lowndes, 57; Lumpkin, 156; Marion, 72; McIntosh, 36; Meriwether, 72, 88, 115; Monroe, 115, 125; Montgomery, 36, 57; Morgan, 152; Murray, 113; Muscogee, 72, 115; Newton, 130; Oglethorpe, 62, 110, 113; Polk, 82, 106; Putnam, 57, 59, 61, 62, 78, 97, 115, 130, 162; Rabun, 39; Randolph, 72; Richmond, 48, 50; Screven, 57; Stewart, 72, 199, 200; Sumter, 72; Talbot, 41, 72; Troup, 41, 64, 71, 72; Twiggs, 79, 100; Upson, 160; Walker, 44, 65, 94; Walton, 39, 152; Washington, 33, 35, 36; Wayne, 39, 40; Wilkes, 3, 11, 41, 48, 54, 63, 67, 74, 111, 115, 130, 151

Georgia Citizen, 121, 153
Georgia Cotton Planters Convention, 119, 153
Georgia Journal, 57, 58
Georgia Messenger (Macon), 173
German servants, 4
Gesner, William, 83
Gibbons, John, 29
Gilmer, George R., 88, 218n
Gilreath, George, 88
Goats, 30; Angoras, 142-145; Asiatic, 142
Godey's Lady's Book, 80, 197
Goodale, Thomas, 24
Grady, Henry W., 224n
Graham, John, 9
Grapes, 159-163; French, 23, 159, 164; Italian, 159; Portuguese, 159
Grasses and clover, 127-134; Alfalfa, 134 (lucerne), 127; Bermuda, 129-131; Italian rye, 127, 128, Kentucky, 127; Oat, 128; Orchard, 134; Rescue, 128-129; Red clover, 129, 134; Tarleton, 128, Terrell, 128; Timothy, 127; Virginia lyme, 127, 128
Graves, John W., 116
Green, E. W., 165
Green, James, 165
Greene, Gen. Nathanael, 32, 52
Gregg, William, 90, 91
Grimes, Thomas, 100
Guano, 187-190; Peruvian, 187

Habersham, James, 10
Hall, Mrs. E. B. L., 115
Hall, Waddell, 157
Hammond, Amos W., 79
Hammond, James H., 215n
Hancock Advertiser, 57, 74
Hardwick, Richard S., 75, 78, 97-99, 112, 115
Hargreaves, James, 52n
Harmon, Garland D., 89, 96, 100-109, 147, 196, 197, 202, 203
Harris, Judge Iverson L., 117, 160, 162
Harris, Miles G., 191
Hart, J. B., 163
Haupt, William, 143
Hemp, 20, 79; Oriental, 47
Herbemont, Nicholas, 162
Hillhouse, David P., 63, 74, 97
Hogs, 30, 145-147
Horizonal plowing, 97-101
Horse-Hoe Husbandry (Tull), 24
Horses, 138-139
Horsley, James, 160

Horticulture, 149-167; Spanish, 149, 160
Horton, Major, 19
Howard, Charles W., 65, 80, 88, 89, 118, 119, 120, 122, 123, 131, 132, 133, 134, 135, 137, 139, 141, 178, 181, 198
Hubbard, Joseph, 136
Huger, John, 162
Hume, Professor William, 163
Hunt, Henry, 162

Indentured servants, 2-5
Indigo, 18-20; Guatamalan, 19
Iverson, B. V., 128

Jackson, Gen. Thomas J., 108
Jefferson, Thomas, 48
Jekyl Island, 19, 150
Johnston, Richard Malcolm, 112, 113
Jones, Dr. Alexander, 58
Jones, James V., 118, 147
Jones, James W., 119
Jones, Jethro, 95, 124, 139
Jones, Joseph, 116, 118
Jones, W. W., 119
Journal and Courier (Savannah), 191
Journal and Messenger (Macon), 63, 126

Keiffer, Theodore, 9
Kentucky grass, 127; mules, 137, 138
Kidd, George, 175
King, Ralph, 66
King, Roswell, 196
King, Thomas Butler, 118
Kollock, George J., 87

La Taste, Victor, 163
Land cessions, 1, 8, 9, 33, 40, 41, 44
Land speculation, 35-38
Land system, Colonial, 5-8, 10-12; of 1784, 33; after 1803, 38-40
Landscape gardening, 173-178
Landrum, William, 143, 224n
Lane, Andrew J., 98, 112, 131
Lanman, Charles, 45-46
Leake, Richard, 29
Lee, Daniel, 66, 86, 95, 100, 101, 118, 119, 120, 121, 122, 129, 135, 141, 151, 157, 178, 179, 203
Lee, Gen. Robert E., 87, 109
LeHardy, Camille, 160
Lenoir, W. A., 139
Lewis, David W., 97, 112, 113, 117, 118, 191
Lime kilns, 82
Logan, Francis, 45

INDEX

London News, 193
Longstreet, William, 209n
Longworth, Nicholas, 171
Loomis, I. N., 125
Lottery, 7, 33, 38-41, 44, 60, 61, 62
Louisiana, 71, 72, 92, 106, 107, 176, 196, 203
Lucas, Eliza, 18
Lumber, 81, 82
Lumpkin, William, 68, 69
Lumpkin, Joseph H., 124
Lyell, Sir Charles, 62, 64, 66, 69, 70, 152

MacDonald, Charles J., 113
Macon Telegraph, 56, 130
Mallet, Gideon, 28
Manifee, George, 58
Manufacturing, 89-92
Martin, Carlyle P. B., 112, 124, 125
Martin, John T., 113
Maryland, 42, 110, 122
Massachusetts, 136
Mathews, Governor George, 36, 37
Matthews, Jacob, 29
McCall, Thomas, 58, 159, 160, 162
McCay, Professor Charles, 77
McDonald, Alexander, 64
McQueen, John, 49
Mechanics Institute of Georgia, 118
Melish, John, 49
Milledge, John, 29, 38, 48
Milledgeville Southron, 59
Miller, Phineas, 53
Miller, Robert, 18
Mims, Dr. Philip, 113
Minerals: coal, 81; copper, 81; gold, 81; iron, 81, 91, 92; kaolin, 81, 82, 183; limestone, 81, 82; marble, 81; saltpetre, 81; slate, 82
Mississippi, 68, 69, 71, 72, 85, 96, 98, 104, 105, 106, 134, 144, 153, 176, 196, 203
Mohawk Valley, N. Y., 155
Monroe, President James, 162
Moody, John W., 113
Moore, N. B., 127
Moses, Raphael J., 154, 155, 166
Mules, 56, 137, 138
Muscogee Democrat, 170

Nacoochee Valley, 45, 136
Naval stores, 24, 25, 81, 82
Neal, David, 162
Nelson, Robert, 80, 83, 86, 117, 121, 152, 153, 154, 155, 169, 170, 172
New Orleans Delta, 83

241

New York Daily Tribune, 155
Nisbett, James A., 117
Nunez, J. M., 163
Nurseries, 158, 159

Oglethorpe, Gen. James Edward, 3, 13, 22, 27, 28, 30
Oglethorpe University, 122
Ohio, 78, 85, 127, 145, 164, 165, 171, 199
Olmsted, Frederick Law, 181
Ornamental trees, 153
Oxen, 137

Parramore's Hill, 62, 152
Peabody, Charles A., 67, 80, 118, 120, 121, 124, 153, 156, 169, 170, 171, 172, 174, 175, 178, 180, 181, 182, 196, 203
Peaches, 55, 150-155
Peanuts, 146, 225n
Pears, 13, 158, 159
Pendleton, Dr. Edward Monroe, 81, 113, 131, 188, 196
Pennsylvania, 21, 49, 50, 69, 85, 127, 132, 134, 147, 157, 182
Perryman, Harmon, 162
Peters, Richard, 69, 83, 84, 101, 118, 125, 131, 133, 134, 135, 136, 138, 141, 142, 143, 144, 147, 148, 155, 158, 203
Philips, Martin W., 101, 104, 105, 134, 201
Pierce, Bishop George F., 112, 113
Pine Mountain Valley, 65
Pitts, John W., 130, 138, 146
Plows, 24, 55, 93-97
Ponce, Dimos, 88, 112, 191, 193, 194, 227n, 230n
Pope, S. L., 113
Potash, 25
Potatoes, Irish, 168-169; sweet, 21
Poultry, 30, 148
Pratt, Daniel, 91
Printup, S. W., 163
Prior, Andrew J., 70

Randall, Henry S., 139
Randolph, Thomas Mann, 97
Redmond, Dennis, 82, 83, 84, 118, 147, 163, 182, 228n
Reese, Dr. D. A., 59
Rheney, John W., 131
Rice, 2, 17-18, 47
Robinson, Solon, 61, 88, 150, 176, 203
Robinson, William P., 158
Rogers, Alphonso, 124
Rome Courier, 101, 194
Rose, Simri, 117, 173

Royall, Mrs. Anne, 42
Rucker, James, 87
Ruffin, Edmund, 91, 107, 109

St. Simons Island, 51, 150, 159
Sapelo Island, 51, 59, 129, 186
Sasnett, Richard P., 76
Savannah Georgian, 59
Schley, Henry J., 140, 141
Schley, Gov. William, 111, 118, 163
Scotch servants, 4
Scott, Thomas F., 123, 199
Sheep, 30, 139-141; English, 141, 223n
Shelburne, Earl of, 162
Shorter, James, 36
Silk, 5, 13-17, 75, 111
Singleton, Richard, 141
Slavery, 2, 5, 196-203
Smith, Thomas J., 189
Soil exhaustion, 61, 72
Sorghum, 83, 86
South Carolina, 5, 8, 18, 27, 28, 45, 48, 53, 56, 69, 79, 90, 91, 110, 117, 141, 142, 144, 164, 169
Southern Literary Companion, 121
Southern Recorder, 65
Soybeans, 86, 216n
Spalding, Thomas, 51, 59, 90, 110, 129, 130, 140, 186
Sparks, Thomas H., 101, 104, 107
Steed, Alexander, 50
Stephens, Alexander H., 181
Stephens, Linton, 112, 113
Stephens, Thomas, 27
Stephens, William, 21, 23, 24, 51
Stocks, Thomas, 118
Strawberries, 170, 172
Stuart, James, 68
Sugar cane, 49, 57

Tailfer, Patrick, 22
Talissee strip, 40
Tallulah Falls, 45
Tariff, 58, 59, 73
Tattnall, Josiah, 38
Taylor, Jeremiah, 156
Tennessee, 44, 56, 59, 77, 82, 117, 137, 138, 139, 144, 145
Terrell, William, 87, 90, 112, 113, 118, 125, 126, 127, 128
Tesnatee Gap, 45
Texas, 66, 70, 71, 72, 84, 98, 121, 127, 143, 144, 152
Thomas, Alexander, 88
Thomas, Gen. Bryan, 109
Thomas, James, 112, 113, 127, 191, 201
Thomas, John S., 118

Thurmond, William H., 118, 156, 158
Ticknor, Francis Orray, 71, 131, 157, 158
Tobacco, 49, 50
Toombs, Robert, 118, 220n
Tray Mountain, 136, 137
Treutlen, John Adam, 9
Trott, John, 64
Trustees, 2, 3, 4, 5, 7, 9, 13, 16, 20, 25, 26, 27, 28, 29
Tull, Jethro, 24
Turner, Thomas M., 191

Uchee Indians, 28
Uncle Tom's Cabin (Stowe), 193
University of Georgia, 77, 112, 122, 125, 150, 172

Van Buren, Jarvis, 83, 94, 118, 127, 151, 155, 156, 157, 158, 163, 166, 188, 203
Vandiver, Adam P., 45, 46
Vann's Valley, 101, 104, 107
Vegetables, 55, 168-172
Virginia, 9, 42, 50, 54, 61, 69, 75, 97, 108, 109, 177

Walker, George A. B., 172
Ward, Malthus A., 150
Waring, George, 127, 137, 139
Warren, L. C., 138
Warren, Williams, 162
Watts, James W., 140, 141
Wayne, Anthony, 32
West Indies, 5, 51, 193
White, Dr. William N., 118, 120, 156, 176
Whitefield, Rev. George, 23
Whitney, Eli, 52, 53
Whitten, Isaac, 191
Williams, Edward, 136, 137
Williams, Edwin, 127
Williams, Robert, 30
Wilson, Dr. John S., 80, 197
Wilson, Lawrence, 165
Wilson, Woodrow, 199
Wine industry, 163-166
Winn, John, 166
Wiregrass Reporter, 102
Witcher, Lacy, 70
Woman's Home Book of Health, 197
Woolfolk, John, 198
Wright, Governor James, 9, 10
Wyly, James R., 113

Yamacraw Indians, 1
Yancy, William L., 220n
Yazoo Land Fraud, 36, 37, 51
Yost, George W. N., 96
Young, Brigham, 144

www.ingramcontent.com/pod-product-compliance
Lightning Source LLC
Chambersburg PA
CBHW030134240426
43672CB00005B/129